P9-CCV-359

Jab, Jab, Jab, RIGHT HOOK

ALSO BY GARY VAYNERCHUK

Crush It!

The Thank You Economy

GARY VAYNERCHUK

Jab, Jab, Jab, RIGHT HOOK

HOW TO TELL YOUR STORY IN A NOISY SOCIAL WORLD

HARPER
BUSINESS

An Imprint of HarperCollins*Publishers*
www.harpercollins.com

Products, pictures, trademarks, and trademarked names are used throughout this book to describe various proprietary products that are owned by third parties. No endorsement of the information contained in this book is given by the owners of such products and trademarks, and no endorsement is implied by the inclusion of product, pictures, or trademarks in this book.

HarperCollins books may be purchased for educational, business, or sales promotional use. For information, please write: Special Markets Department, HarperCollins Publishers, 10 East 53rd Street, New York, NY 10022.

FIRST EDITION

Image of boxing gloves courtesy of miflippo/istockphoto.

Library of Congress Cataloging-in-Publication Data has been applied for.

ISBN 978-0-06-227306-2 (Hardcover)

ISBN 978-0-06-231671-4 (Signed Edition)

13 14 15 16 17 OV/QGT 10 9 8 7 6 5 4 3 2 1

TO MY TWO KIDS, MISHA AND XANDER. YOU HAVE
BROUGHT A KNOCKOUT PUNCH OF LOVE TO MY LIFE
THAT I DIDN'T KNOW EXISTED.
AND TO THE WOMAN WHO BROUGHT THEM TO ME,
THE LOVE OF MY LIFE, LIZZIE.

CONTENTS

ACKNOWLEDGMENTS

I have so many people to thank I could never fit all their names into a tweet, so I decided to put them in an acknowledgments page.

First and foremost, I want to thank my family, whom I love very much and who always help me, support me, and drive me. They really are the guiding light to my life.

I also want to thank Stephanie Land, who is my true partner in writing these books. This is the third book we've written together. Steph—thank you so much. I truly, truly could never write a book without you.

Huge shout-out to my boy Nathan Scherotter, who is the CEO of this book. Nate has been an amazing friend and business associate for many years. His help in guiding this book's content and then its sales afterward was immensely important. I love him like a brother—except when we play basketball against each other.

I'd also like to thank everyone at VaynerMedia who helped with this project. Kelly McCarthy, Marcus Krzastek, and Etan Bednarsh—thank you so much for being such great partners and family. Another big shout-out to the crew of Vikash Shah, Steve Unwin, Sam Taggart, Colin Reilly, Alan Hui-Bon-Hoa, Haley Schattner, India Kieser, Jed Greenwald, Jeff Worrall, Katie Katherine Beattie, Nik Bando, Patrick Clapp, Michael Roma, and Simon Yi for helping with the content of

#JJJRH. Also to Andrew Linfoot, George Barton, and Kyle Rosen for interning over the summer and helping with the process.

Another big thank-you to everyone at HarperCollins. Hollis Heimbouch and her team have always been great to work with and a valuable asset every step of the way.

But most of all, I want to thank all the fans and others who have been following for the past four or five years as I've discussed the current trends. Obviously, it is a cliché to say I wouldn't be here without you—but the fact is, I wouldn't. If you didn't continue to buy my books, read them, or react to them, I wouldn't write them. This is for you.

Last, and as always, I have to say thanks to my immediate family, whom I love with all my heart. My parents, Sasha and Tamara, and my grandmother Esther. To my brother A. J. and his wonderful girlfriend, Ali. To my sister Liz, my brother-in-law Justin, and their two kids, Hannah and Max. To my brother-in-law Alex, his wife, Sandy Klein, and their children Zach and Dylan. As well as to my wonderful in-laws, Peter and Anne Klein. All of you mean the world to me.

AUTHOR'S NOTE

At the time of this writing, I hold Facebook stock. I also hold Twitter stock, bought in 2009. I owned Tumblr stock that was sold in the 2013 sale to Yahoo.

I don't hold any Snapchat or Pinterest stock, but I wish I did.

I have been careful not to critique any competitors of current VaynerMedia clients for the case studies that appear in this book.

INTRODUCTION: WEIGH-IN

A look at my Twitter feed during football season reveals that pretty much the only thing that can dampen my optimistic outlook and love for life is when the New York Jets do something stupid, like their quarterback running into his offensive lineman's ass and fumbling the ball, giving the opposing team a touchdown. You know, the usual. It's no secret that I intend to buy them one day. Maybe not from Woody Johnson, maybe from his successor, but someday. So every loss cuts me to the quick. Yet while my heart belongs to football, that's not the sport that dominates my thinking most of the time. Most of the time, unless I'm with my family, I'm doing business. Which means that most of the time, like a lot of other businesspeople, marketers, and entrepreneurs, I'm boxing.

Fast-paced, competitive, and aggressive, boxing is a natural metaphor for doing business. And despite its decline in popularity over the last few decades, we have incorporated its lingo into our language probably more than any other sport. I hear it in boardrooms all the time. When managers and marketers outline their social media strategies, they often talk about the "knockout punch" or "right hook"—their next highly anticipated sale or campaign—that's going to put the competition out for the count. Their eyes gleam the same way a twenty-year-old Mike Tyson's probably did right before it took him less than six minutes to knock out Trevor Berbick and become the youngest heavyweight champion in boxing history. They're bloodthirsty. Even at companies where I see impressive efforts to patiently build the relationships so crucial to successful social media campaigns, marketers are itching to land the powerful, bruising swing that will knock out their opponent or their customer's resistance in one tooth-spritzing, killer blow. Right hooks, after all, convert traffic to sales. Right hooks earn Cannes Cyber Lions Awards. They easily show results and ROI. Except when they don't.

That's the truth, isn't it? We've seen a few well-placed social media hits over the years, but more often than not, social media marketers are throwing their best right hooks all over Facebook, Twitter, Instagram, and YouTube, yet still failing to land killer blows in the form of increased sales and market share. They're swinging as hard as they can, and then . . . whoosh. No connection. And it's not that the spot didn't reach any eyeballs. People saw it; they just didn't care. Despite terrific customer awareness, the brand's content wasn't compelling enough to inspire consumers to do anything with it.

I thought it would be three or four years before I wrote another book. I thought I'd said everything I needed to say for now. I've been on a mission to convince marketers that today, business is all about making the customer happy. After spending so much time preaching the importance of the jab—the one conversation, one engagement at a time that slowly but authentically builds relationships between brands and customers—the last thing I wanted to do was write a book that outlines how to execute a killer right hook with content. Because I suspect that deep down inside, if given the choice most businesspeople would ditch the whole social engagement thing and just go for the punch, because the social engagement thing is hard and takes too much time. We're primed for immediate gratification, and if we don't have to be patient, we won't. So I'm nervous that if I put out a book offering a blueprint to the perfect content for every major social media platform relevant today, a lot of people are going to think they can now ease up on the time-consuming task of engaging with their customers. Armed with a foolproof, knockout right hook, you don't need as many jabs to win, right?

Wrong. Wrong, wrong, wrong, wrong, wrong.

There's a reason why boxing is called "the sweet science." Critics dismiss the sport as mindlessly barbaric, but where they see violence, those of us who understand and respect it see strategy. In fact, boxing is often compared to chess for the amount of strategic thinking it requires. The right hook gets all the credit for the win, but it's the ring movement and the series of well-planned jabs that come before it that set you up for success. Without a proper combination of jabs to guide your customer—I mean, your opponent—right where you want him, your right hook could be perfect and your opponent could still dodge it as easily as a piece of dandelion fluff. Precede that perfectly executed right hook with a combination of targeted, strategic jabs, however, and you will rarely miss.

The realization that I had to write this book occurred in late 2012, on a red-eye flight home from the West Coast. I was exhausted, slumped against the side of the plane with my forehead pressed against the window because I was too tired to hold my head up. And I was thinking back to Wine Library TV, the online wine video blog that launched my career in social media marketing and helped pave the way to where I am now.* I've always credited the success of that venture to my hustle and single-minded dedication to engaging with my fans and customers by answering every email or blog comment I received and going overboard to show my appreciation for their business. But I had just spent another day analyzing a potential client's floundering, misguided, and downright lackluster social media campaign. Despite their earnest efforts to engage with their customers, they were seeing little brand awareness or sales momentum. And as I sat there mulling over how to help them, not sure whether I was about to catch my second wind and answer emails, or drop into a coma, I had an epiphany. The content. When I launched Wine Library TV, I chose to do long-form video blogs, about twenty minutes apiece, on a platform (YouTube, and later in 2007, Viddler) where asking people to stick around for five minutes was like asking them to sit through the des-

* I ended Wine Library TV on its one thousandth episode. Thanks to everyone who asked me to bring it back, in particular @StanTheWineMan.

ert scene in the uncut version of *Lawrence of Arabia*. And yet they had stuck around, relaxing with their feet up in front of their computers, to watch me taste wine and hear my opinions. Why? Maybe Wine Library TV didn't catch fire just because I hustled more than everybody else. Maybe its popularity wasn't just due to my unique combination of expertise, humor, and irreverence (not to mention spellbinding charisma). My high-quality content definitely factored in, but that might not have mattered had I not also made native content—authentic content perfectly crafted for that particular new platform, YouTube, not thanks to good lighting or smart editing, but because it embraced authenticity and "realness." And maybe I needed to make sure that my clients and others who turned to me for advice were doing the same.

The business world had stubbornly resisted accepting that a short-term approach to social media wasn't going to work, so I'd spent the majority of my time and effort over the years emphasizing the importance of the long view, and teaching people how to communicate in such a way that would develop authentic and active customer relationships. My last book, *The Thank You Economy*,* could easily have been titled *Jab, Jab, Jab, Jab, Jab!* It was told in two parts: One half built a strong argument for the ROI of jabbing your customers—engaging through incredible, heartfelt service and social media—and the other half presented case studies illustrating great jabs and how they increased conversion rates. But while it's true that you can't land a solid right hook if you don't set up the punch with a series of good jabs, it's also true that no fight has ever been won on jabs alone. Eventually, you have to take your shot. Sitting on that plane, I realized that I had become so intent on perfecting people's jabs, I had neglected to pay enough attention to improving their right hook.

One reason I talked very little about the actual moment of conversion in *The Thank You Economy* was that it came out on the heels of my first book, *Crush It!*,† which explained what great content should look like and introduced a number of platforms that to many seemed bizarre and even pointless at the time but which have now become widely accepted as crucial to business. But that was four years ago. Pinterest and Instagram were still in development. The majority of our Facebook status updates were text, not photographs. No one

* Buy it, it's good.

† Buy that one, too!

owned an iPad. Right hooks have to be done differently now, thanks to the massive change and proliferation we've seen in social media platforms. I wasn't sure I wanted to write another book, but I had to, because what I've learned in the past year or so is so pressing it needs to be said right now. I think I know what the future of successful marketing looks like. What else is new? As usual, a lot of people will disagree with me. But I think I'm right, and I like that feeling.

Jab, Jab, Jab, Right Hook is an update of everything my team at VaynerMedia and I have learned about successful social media and digital marketing through the work we've done with thousands of start-ups, Fortune 500 companies, many celebrities, and a substantial number of entrepreneurs and small businesses since that day on the airplane. A mash-up of the best elements of *Crush It!* and *The Thank You Economy* with a modern-day spin, it offers a formula for developing social media marketing strategies and creative that really works. We'll still discuss engagement, because I still think most people aren't engaging enough to set up their jabs as well as they should, but this book will emphasize right hooks. Specifically, how to create perfect and distinct native content for every one of the multiple platforms you now have to use to cross-pollinate your brand and message.

No matter who you are or what kind of company or organization you work for, your number-one job is to tell your story to the consumer wherever they are, and preferably at the moment they are deciding to make a purchase. For a long time, we did that through television, radio, and print. We evolved with the times, eventually attempting guerrilla marketing, sending emails, and creating banner ads. But the attention-grabbing power of these older platforms is weakening, their audience is shrinking, and every day that goes by costs us more to reach fewer people, because though those older platforms still serve their purpose, people are just not watching television, listening to radio, reading print, or even paying much attention to emails. At least, not as often as they used to. They're on social media.

These platforms still feel new and untested. I get that. But enough waiting around. Now that the infrastructure is built and the plumbing is in place, it's time to learn how to use the system to achieve your business objectives, and to put more time, energy, and dollars into the place where the consumers actually are, and not where you wish they would stay. Social media platforms offer us our best chance to stretch our working dollars the furthest.

Consider this book a training camp to prepare you to storytell on today's most important social media sites. In order to ensure that the book retains its value over time, all of the platforms selected for analysis have at least three to five years left in them (that's actually an impressive life span for an Internet platform). You'll learn how to create the storytelling formula that will most resonate with consumers as they look at their mobile device forty times a day. In addition, we will examine good, bad, and ugly examples of some famous and not-so-famous brands' social media storytelling. In this way, I hope to deliver on the promise I made myself when I decided to write this book: to create a guide to steer people away from common social media marketing pitfalls, and a reference that people can come back to again and again. As in boxing, once you learn the science of the social media sport, you will be able to apply what you learn in these rings to any platform that crops up in the future. And that's a great story.

I see this as the final book in a trilogy that covers not only the evolution of social media, but my own evolution as a marketer and businessman as well. (My next book will probably be about parenting. Or maybe root beer. Or even how I bought the Jets.) The world changes, platforms change, and we learn to adapt. But the secret sauce remains the same: The incredible brand awareness and bottom-line profits achievable through social media marketing require hustle, heart, sincerity, constant engagement, long-term commitment, and most of all, artful and strategic storytelling. Don't ever forget it, no matter what you learn here.*

* Please.

Jab, Jab, Jab, RIGHT HOOK

ROUND 1: The SETUP

Where's your phone?

In your back pocket? On the table in front of you? In your hands because you're using it to read this book? It's probably somewhere within easy reach, unless you're one of those people who are constantly misplacing their phones and my question has you rummaging through the laundry basket again or checking under your car seat.

If you're in a public space, look around. I mean it, pop your head up. What do you see? Phones. Some people are doing the old-fashioned thing and using them to actually talk to another person. But I predict that someone, and probably several someones, within a four-foot radius is playing Dots. Or double-tapping a picture. Or composing a status update. Or sharing a picture. Or tweeting. In fact, unless you're visiting Aunt Sally in the nursing home—and even then, you'd be surprised at how iPads are crashing the ninety-year-old demo lately—it's more than likely that almost everyone around you has a smartphone in his or her possession, and if not a phone, then a tablet. I know this because there are nearly 325 million mobile subscriptions in the United States alone.

And when people are using their devices, it's probable that almost half are networking on social media.

If I wrote that line correctly, it should read with the kind of serious tone we reserve for Very Important News. But what's the big deal? By now everyone gets it—social media is everywhere. It has changed the way society lives and communicates. It's no longer just the first adopters and the young who are hooked—71 percent of people in the United States are on Facebook, more than a half billion globally are on Twitter—a population that includes everyone from the pope to a parrot named Rudy, and almost every small business in America in between—and almost half of all social network users check in on these sites at least once a day, often as soon as they wake up in the morning. It has altered the way people fall in and out of relationships, stay in touch with family, and find jobs. Finally, there are few if any holdouts who will deny that today business simply can't be done without it, especially when one in four people say they use social media sites to inform their purchasing decisions.* Boomers, who control 70 percent of U.S. spending, increased their social usage 42 percent in one year. Moms, the buyers and budget analysts for most families, are all over it. The eyeballs marketers want to reach, the ones belonging to people who make purchasing decisions and who have money to spend, are spending increasing amounts of time on social media sites. They are doing this because they are no longer tied to their laptops and PCs to get their social media fix. Thanks to their smartphones and tablets—and eventually, their glasses and who knows what else—where they go, their social networks go, too.

Social media is like crack—immediately gratifying and hugely addictive. With their mobile devices in hand, people may as well be getting intravenous drips of the stuff, a constant and incredibly noisy stream of information, imagery, and interaction. And as with any drug (so I'm told—dead serious, I've never tried anything), the more they get, the more they want. That's why it matters that more than half the total U.S. mobile population are using their mobile devices to engage on social media sites—they're there so much that it's starting to alter the way they want to interact with brands, services, and businesses, even when they're *not* on social media sites.

Very Important News? You bet your ass it is.

* And that's for now, only about seven years into the social media phenomenon. In five years, it will probably be one in two people.

HOW SOCIAL BLENDED INTO DIGITAL

This statistic alters current fundamental marketing principles. Over the last half decade, marketers have learned to divide their campaigns into three categories—traditional, digital, and social. We knew that traditional marketing had begun to lose much of its relevance and reach with the advent of the Internet and digital media options pulling the audience away from television commercials and print. Still, when properly aligned, these three platforms could often complement each other effectively. But now that people are addicted to their social networks, they get itchy when their media experience doesn't have a social element, and they move on. Social media is no longer just pulling the audience away from traditional marketing; it's cannibalizing digital media, too.

The evidence is clear. Emails, banner ads,

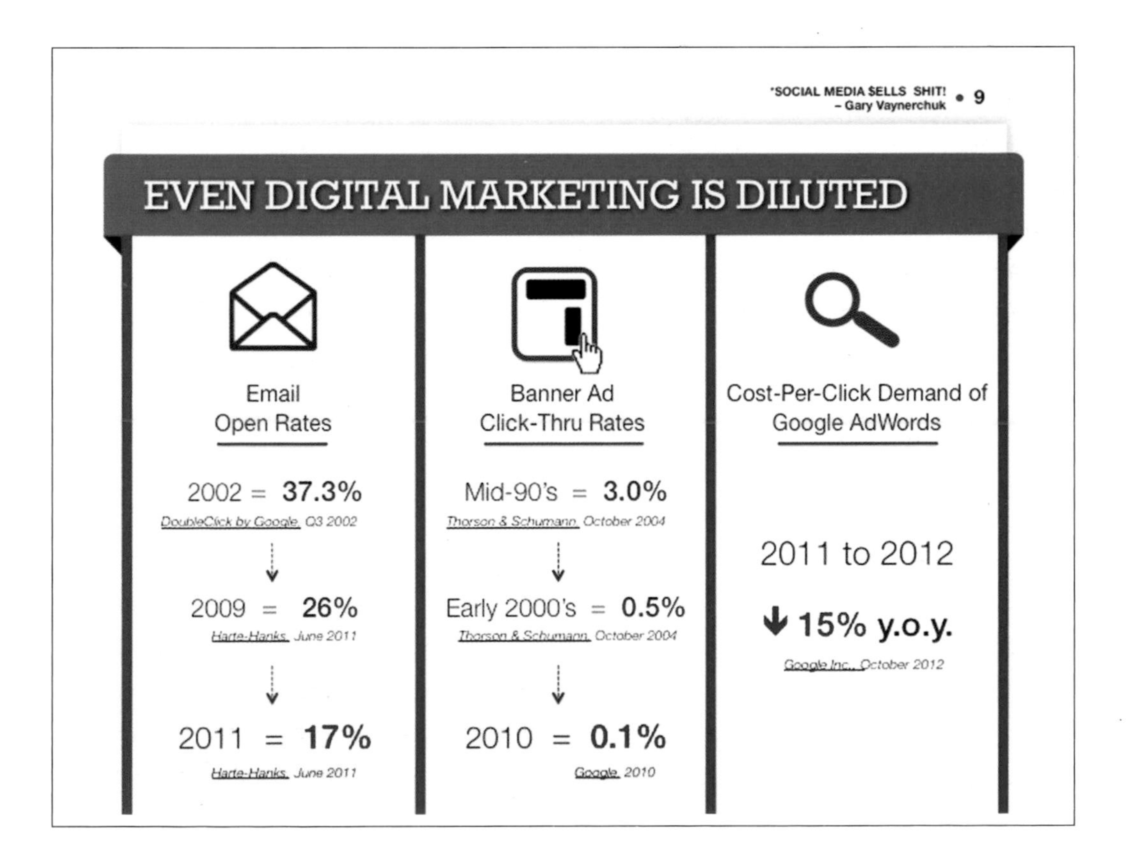

search engine optimization (SEO)—the power of all these stalwart digital marketing tactics of the Internet era is diminishing, with one exception: when the digital platform has a social media component. In fact, adding a social layer to any platform immediately increases its effectiveness.

Anyone who pays attention to media trends and history shouldn't be surprised. It's natural that every new marketing platform would usurp the one that came before. Radio leached away the audience for print, TV poached the audience for radio, the Internet stole audience from every one of these old platforms, and now social media (which is really just the evolution of the Internet) is well on its way to overtaking them all. What's astounding, however, even to me, is the speed at which this progression is happening. It took thirty-eight years before 50 million people gained access to radios. It took television thirteen years to earn an audience that size. It took Instagram a year and a half.

With the instant access to social media made possible by mobile devices, there's no such thing as undivided attention anymore.* It's not just that people are scrolling Facebook on their laptops while hanging out on the couch half watching *The Voice*. They're sharing on Pinterest while crossing the street. They're loading on Instagram while driving. And while they're tweeting at the supermarket, they're starting to ignore the expensive endcaps brands paid for at the end of the aisles, as well as the displays of candy and magazines in front of the registers.† From a personal safety point of view, mobile social networks are a disaster—no one is looking where they're going. But from a marketing perspective, the writing is on the wall: The fastest-growing marketing sector getting people's attention is social media. The strict dividing lines between marketing categories can no longer exist—they must all be blanketed with a layer of social.

The problem is, most companies, marketers, and entrepreneurs haven't gotten the message, and so they persist in overpaying for diminishing returns.

It's not that businesses aren't trying. Many were dragged kicking and screaming to social media, but by now most understand that having a Facebook page and a Twitter account is critical for brand visibility and credibility. So they're there. They're just still not doing it right. While compa-

* A lot of people are wailing and gnashing their teeth about this. That's evolution, baby. Get over it.

† *Ad Age*, when you finally recognize this major trend, please credit *JJJRH* for being the first to point it out.

nies were getting comfy cozy with the idea of being on social media platforms, social media transcended those platforms, and few businesses have followed.

Marketers and business leaders have got to catch up. People want to be social wherever they consume their media. This means that you need to fold a social element into all of your creative, including for traditional media, and into every interaction with your customers, whether by commenting on Tumblr, gamifying a banner ad, engaging on a news aggregator, or sending people to Facebook at the end of your thirty-second radio spot. From now on, every platform should be treated as a social networking platform.

And now that your consumer is mobile, you'd better be, too.

A quick look at many companies' marketing efforts reveals that many have caught on that mobile networks and apps present the biggest opportunity for brand growth. They are disseminating content all across the mobile social media board, making their presence known on all of the most popular networks, like Facebook, Twitter, Instagram, Pinterest, and Tumblr. For the most part, their content looks like this:

With the exception of the Twitter feed, can you tell which platform is which? Though some platforms may eventually implement changes that might alter this scenario by the time this book goes to press, as of this writing, you can't.

I write this with the utmost respect: Marketers, small businesses, celebrities, I know you're trying, but with a few exceptions, the content you're putting out there sucks. You know why? Because even though consumers are now spending 10 percent of their time with mobile (a number that is soon going to be much higher), you're investing only 1 percent of your ad budget there. You can't just repurpose old material created for one platform, throw it up on another one, and then be surprised when everyone yawns in your face. No one would ever think it was a good idea to use a print ad for a television commercial, or confuse a banner ad for a radio spot. Like their traditional media platform cousins, every social media platform has its own language. Yet most of you haven't bothered to learn it. Most big companies haven't put in the financial resources, and most small businesses and celebrities aren't putting in the time. You're like tourists in Oslo who haven't bothered to study a word of Norwegian. How can you expect anyone to care what you have to say?

Whether you're an entrepreneur, a small business, or a Fortune 500 company, great marketing is all about telling your story in such a way that it compels people to buy what you are selling. That's a constant. What's always in flux, especially in this noisy, mobile world, is how, when, and where the story gets told, and even who gets to tell all of it.

This book will show you how to create the kind of shareable, relevant, value-driven content that ensures consumers always pay attention to your story, no matter where they go, and then that they pass on your content, creating the word of mouth critical to actually making the sale. Ultimately, that's the real reason to do any of this—because social media sells shit.

HOW STORYTELLING IS LIKE BOXING

Until recently, traditional marketing was nothing but a one-sided boxing match, with businesses slamming right hooks onto the same three or four platforms—radio, television, print, outdoor, and then later, the Internet—as fast and as often as possible.

"Two for one, today only!" Punch.

"Grab your keys and come on in!" Punch.

"Don't miss this once-in-a-lifetime opportunity!" Punch.

It was an unfair fight, but it worked. Customers had to take the hit since they had nowhere else to go to consume their media. Social media, however, finally gave them an advantage. Now the match was taking place on a platform that allowed them to demand a change in how the game was played. They were going to demand more time. They wanted their brands and companies to spar with them a little, pay attention to them, let them voice their opinions and concerns, and make the brand their own before giving them a shot at the hard sell. From now on, marketers were going to have to spend a lot more time jabbing at their consumer before landing their right hook.

That's why I spent the majority of my last two books explaining how to jab properly, even though I knew that managers and marketers cared mostly about right hooks. Jabs are the lightweight pieces of content that benefit your customers by making them laugh, snicker, ponder, play a game, feel appreciated, or escape; right hooks are calls to action that benefit your businesses. It's just like when you're telling a good story—the punch line or climax has no power without the exposition and action that come before it. There is no sale without the story; no knockout without the setup.

Ironically, over the past few years, the same technology that made it possible for marketers to successfully jab—to use social media to tell their story by engaging directly with their customers—has also made it ten times harder to actually reach those customers and convert the sale. Even the businesses that got in on social media early are now seeing diminishing returns on some of their efforts. While they're working to get those jabs just right (and there is still room for a lot of improvement), companies also need to update and improve their right hook techniques. They need to pay attention to context. They need to think about timing. They need to start respecting the platforms and understand the nuances that make them interesting.

At the heart of the content quality crisis is the fact that many marketers and small businesses still don't believe in social media or even really understand it. They have a presence on social media platforms, but only because they realized they had to in order to be taken seriously. Though the interaction required by social media is like oxygen and sunshine to people like me, and others who have built successful businesses through these platforms, many marketers remain skeptical. Publicly, they claim to be thrilled to have the opportunity to engage directly with their customers; privately, they suspect, maybe even fervently hope, that Facebook and its spawn are fads. Because things were a heck of a lot easier before social media. If you were a big business you created a campaign, like the Geico cavemen, plastered it as far and as wide as possible, and sat back to see what happened. You used the same images and ideas for television, print, and outdoor. If the reports showed the campaign didn't work, you blamed the data collection technique or some other random element. After six months, regardless of whether the campaign worked or not, you scrapped it and started over with a brand-new one. If you were a small business, you sent some fliers in the mail, created a cute little Yellow Pages ad, ran a local radio ad, and waited for people to come in. If you were really forward-thinking in the late first decade of the 2000s, you did some SEO! Wow!

Now, if you truly understand how marketing works today, you know there is no individual six-month campaign; there's only the 365-day campaign, during which you produce new content daily. Maybe you come up with three big campaign ideas—if you're Geico, it might be the gecko, Maxwell the Pig, and Dikembe Mutombo happily blocking the shot—but you run them simultaneously, selecting a different platform for each, and only using the one that gets the strongest response as the seed for a television ad. Now, if you know what you're doing, you scour the Internet daily, searching for references to your product or service so you can jump in on the conversation, or scrambling at a moment's notice to respond to a 2:47 P.M. complaint via Twitter. To do social media right is harder and requires more time and effort than most people realize. And though the analytics get more accurate and sophisticated by the day, even the best right hooks can sometimes take a while to offer quantifiable, data-driven proof that they worked (like when you post call-to-action driven content asking people to buy airline tickets or a bottle of wine). So though the majority of marketers and businesspeople are working with social media, a lot of them are still questioning the value of the platforms, and few respect them enough to fully invest, either financially or philosophically. It shows. It shows in the low frequency of their posts, the inferior quality of their content, the lack of ingenuity with which they approach each new medium even as it gains in popularity, and worst of all, in the shocking lack of effort put toward showing care and respect for any community that has formed around their business despite all the previously listed failings.

Here's how most marketers react to new platform: Someone emails them an article that says something like Snapchat is exploding, so they head over to the site to see what it's about. They spend a few minutes there and see a bunch of drunken twenty-five-year-olds posting bikini shots and text saying, "Walking the dog!" "Anchovies . . . FTW!" They write the site off as a waste of time and don't come back until twelve months later, when everyone and their aunt is using it, at which point they make a big announcement praising themselves, as though being last in line is something to be proud of: "Look what we did! Isn't it exciting? See how responsive we are?" It's embarrassing. It pisses me off. (It also makes me perversely happy because their cluelessness has a definite upside for my clients, my friends, and for me.)

A smart entrepreneur or open-minded brand manager, however, will head over to a new platform, see the bikini shots, and think, "How can I do better?" He or she will spend twelve months securing a solid dominance over the platform in their category, all the while reaping tons of earned media as bloggers and reporters chronicle their progress and analyze their strategy, and attracting the best young talent because the students coming out of business school want to work at progressive businesses. You'd think, given the advantages, that brands and small businesses would be scrambling to be first to market on these platforms, but most of the time their fear of failure, their legal department's fear of lawsuits, or their perceived lack of time outweighs their sense of possibility. They are playing defense instead of offense.

Here's my dirty secret: Though I get to things early and can often see the future, I'm not Nostradamus. I'm not even Yoda. I'm just the kind of person who shows new platforms the respect they deserve. I won't predict what platform will see 20 million users in a year, but once it feels to me like it will, I put my money and my time there, testing the waters, trying new formulas, until I figure out how to best tell my story in a way the audience for that platform wants to hear.

I can't believe how many marketers will dismiss the media habits of five million people. Just because your teenage daughter and her friends are excited about a new platform does not mean that that platform is irrelevant to you or your brand. You may not see any value in sharing your thoughts on nail polish, or posting a picture every time you get a new tattoo, or telling the world every time you set foot in a Wendy's, but when 20 million other people do, you need to do something with that information. Ignoring platforms that have gained critical mass is a great way to look slow and out-of-touch. Do not cling to nostalgia. Do not put your principles above the reality of the market. Do not be a snob.

You cannot win big in social media if you're going to be afraid of emerging technology. Those of us who spent time on YouTube in 2006 watched more than our fair share of dumbasses putting Mentos in cola or dressing their cats in silly outfits. But like a parent who knows that the infant currently squeezing handfuls of peas into mush will grow up to use a fork and knife, we had faith that this platform hadn't yet reached its full maturity or potential. Some people saw an amateur video distribution site; we saw the future of television. For my part, I experimented and tested ideas to see what worked; I created an opening hook like an old-time radio program to make myself more meme-orable. I approached it as a major platform, and so did a lot of other people who are now big-name brands. (And they didn't make the huge misstep to leave YouTube for Viddler in 2007, leaving millions of free views on the table, like I did. Even I screw up sometimes!) We didn't do anything more than take the platform seriously and put in a massive amount of effort to figure out how to make them work for us, committing to the same intense process of testing and observation as any champion boxer before a fight.

A boxer spends a lot of time analyzing his own technique, but spends an equal amount of time analyzing his competitor's technique, too. Even when two fighters meet in the ring for the very first time, they already know each other well. For months before the match, in addition to their regular predawn training in the gym and practice ring, the competitors spend hundreds of hours studying each other on film. Like insanely fit behavioral scientists, they analyze every move and swing their opponent has made in previous fights, repeatedly rewinding and rewatching footage in an attempt to memorize their opponent's technique, and particularly, the tics and habits that can warn a fighter of the swing that's about to come. Does the opponent blink before he throws with his right hand? Does he hesitate to come back after getting hit with a cross? Does he drop his hands when he gets tired? Finally, on the day of the fight, a boxer will take all of this information into the ring with him, armed with a strategy precisely calibrated to take advantage of his opponent's weaknesses and protect himself from the other's strengths, so he can use his best moves to maneuver himself into a winning position.

If, whenever they approached a platform, more marketers prepared their stories with the same intensity as boxers, they'd create much better content. Like great boxers, great storytellers are observant and self-aware. A great storyteller is keenly attuned to his audience; he knows when to slow down for maximum suspense and when to speed up for comic effect. He can sense when he's losing people's interest and can make adjustments to his tone or even to the story itself to recapture their attention. Online marketing requires the same kind of audience awareness, which we can achieve thanks to the tremendous data mining opportunities at our fingertips. The real-time feedback that social media makes possible allows brands and businesses to test and retest, with scientific precision, what content connects with their audience, and what leaves them cold. Ignoring the deep analytics available for your fan page through Facebook (and through other platforms soon) is the equivalent of stepping into the ring without even having watched a video of your opponent during a fight.

WHAT MAKES A GREAT STORY?

A great marketing story is one that sells stuff. It creates an emotion that makes consumers want to do what you ask them to do. If you're a mobile company, you want to motivate people to sign up for a subscription to your service; if you're Disney, you want to impel people to book flights and hotels and come spend money at your park; if you're a nonprofit, you want to move people to make a donation. Your story isn't powerful enough if all it does is lead the horse to water; it has to inspire the horse to drink, too. On social media, the only story that can achieve that goal is one told with native content.

Native content amps up your story's power. It is crafted to mimic everything that makes a platform attractive and valuable to a consumer—the aesthetics, the design, and the tone. It also offers the same value as the other content that people come to the platform to consume. Email marketing was a form of native content. It worked well during the 1990s because people were already on email; if you told your story natively and provided consumers with something they valued on that platform, you got their attention. And if you jabbed enough to put them in a purchasing mind-set, you converted. The rules are the same now that people spend their time on social media.

It can't tell you what story to tell, but it can inform you how your consumer wants to hear it, when he wants to hear it, and what will most make him want to buy from you. For example, supermarkets or fast-casual restaurants know from radio data that one of the ideal times to run an ad on the radio is around 5:00 P.M., when moms are picking up the kids and deciding what to make for dinner, and even whether they have the energy to cook. Social gives you the same kind of insight. Maybe the data tells you that you should post on Facebook early in the morning before people settle into the workplace, and then again at noon when they're taking a lunch break. The better you learn the psychology and habits of your social media consumers, the better you can tell the right story at the right time. A story is at its best when it's not intrusive, when it brings value to a platform's consumers, and when it fits in as a natural step along the customer's path to making a purchase.

Only you know what your story should actually say. At one point it might be "Our barbecue sauce will win you first prize at the chili cook-off," but later you might decide it's more important to tell the story that "our barbecue sauce has all-natural, locally sourced ingredients." How did MasterCard know the time was right for the "Priceless" campaign? Nike had tried a number of stories before it hit on "Just Do It." There are a number of tropes that often work, but ultimately the story that you decide needs telling can change from day to day, even from hour to hour. The perfect story is spun from your intimate knowledge of your history, your competition's history, and increasingly, what you see going on in the world and what you discover your consumers want to talk about.

Whatever story you tell, you must remain true to your brand. Native storytelling doesn't require you to alter your identity to suit a given platform; your identity remains the same no matter what. I'll behave one way when I'm giving a presentation to a client in Washington, D.C., another way while I'm standing on the train platform waiting to head home, and yet another way when I'm watching football with my friends that night. But I'm always the same guy. Different platforms allow you to highlight different aspects of your brand identity, and each jab you make can tell a different part of your story. Have fun with that. One of the biggest mistakes big brands make is to insist that their tone remain exactly the same no matter what platform they're using. In clinging to this outdated model, they're missing out on one of the greatest benefits of social media—always having more than one option.

Entrepreneurs will have an easier time taking advantage of these options because they aren't bogged down by the same red tape as Fortune 500 companies. While entrepreneurs and start-ups can respond with ease to real-time consumer feedback, corporate companies usually take a long time to steer their big old ships around. Because of their smaller size, entrepreneurs can make decisions quickly. Because they don't have a herd of lawyers analyzing their every word, they can keep their sense of humor. They are able to retain their personality and humanity no matter what platform they're on. Once start-ups grow enough to join the ranks of Corporate America, they often become overly cautious and start sticking to the safest, narrowest lane they can find.

THE SWEET SCIENCE

Marketers are constantly asking me for a fixed storytelling blueprint, something that delineates the optimal number of jabs before it's appropriate to throw a right hook. That blueprint doesn't exist. Social media storytelling is as sweet a science as boxing, requiring constant experimentation and hours of observation. Successful online content marketers pay especially close attention to variables such as environmental fluctuations and demographic shifts. At what times do we see the highest level of response? What happens when we use slang? How does the same image work with different taglines? Did it make a difference to add a hashtag? Is there an increase in engagement when we put out animated GIFs? The answers are out there if you learn how to test properly and correctly interpret the data. You can see right away how many people heart on Instagram; how many fans share and comment on Facebook; who repins on Pinterest and how often; how many people reblog and write notes on Tumblr.

Allocating the time and budget for these analyses can be tough for both small and large businesses, but it is imperative. It's not enough to experiment—you have to respond to what the results tell you. This is how you devise a formula to guide your future storytelling on the platform. But that formula should be treated only as an overarching framework, because like any boxer, you can't use the same move over and over again. A fighter will concentrate on trying to hit his opponent's body if he learns that the competitor is reluctant to get hit there. But the next guy he fights might not be afraid to get hit in the body, so he'll have to change his approach.

Similarly, each platform is unique, and requires a unique formula. What works on Facebook won't necessarily work on Twitter. Stories told through pictures on Instagram don't resonate the same way when told in an identical manner on Pinterest. Posting the same content on Tumblr as on Google+ is the equivalent of the tourist deciding that since he can't speak Norwegian he'll just speak Icelandic and it will do. That's stupid. Both languages share similar roots and are spoken by tall, gorgeous blondes, but aside from that, they're totally different.* Today, getting people to hear your story on social media, and then act on it, requires using a platform's native language, paying attention to context, understanding the nuances and subtle differences that make each platform unique, and adapting your content to match. There is a science to creating memorable, effective social media content for mobile that converts fans into customers. Now is the time to learn it.

Today's perfect right hooks always include three characteristics:

1. They make the call to action simple and easy to understand.
2. They are perfectly crafted for mobile, as well as all digital devices.
3. They respect the nuances of the social network for which you are making the content.

I'll share more information that can help you improve your jabs, but I would like to try to get you to start throwing them in different places than you're accustomed. I used to talk about going where the eyeballs go, but consumers would need sixteen eyeballs apiece to keep up with the multitude of devices and media that compete for their attention now. Every marketer's goal is to reach consumers at the moment when they're most influenced to buy. To do that, you need to be where they are. That's a tough proposition when where they are is changing all the time, but it can be done. No matter where you go to meet your customer, however, you had better show up with a knockout story, and some killer content with which to tell it.

* I hope this sentence increases my sales in Iceland. I've long had this crazy desire to be hugely popular in Iceland.

ROUND 2: The CHARACTERISTICS of GREAT CONTENT and COMPELLING STORIES

The social media revolution wrenched the keys to the cultural kingdom away from pundits and gatekeepers, giving ordinary people a voice. But the sound of so many people talking at the same time—not to mention opining, debating, entertaining, instructing, and doing all the other things people do to make their views known online—is overwhelming. In order to increase their odds of being seen and heard, many marketers respond by posting a constant, steady stream of fresh content to their social networks. But the social media equation requires quantity *and* quality. Far too much of the content businesses and celebrities put out is no more innovative or interesting than a Yellow Pages ad. You can find truckloads of garbage on these platforms, especially when they are young and people are compulsively tossing content around like Mardi Gras

beads, or when they are old and act their age. Brands and small businesses want to look relevant, engaged, and authentic, but when their content is banal and unimaginative, it only makes them look lame. Content for the sake of content is pointless. Tone-deaf posts, especially in the form of come-ons and promos, just take up space, and are justifiably ignored by most of the public. Only outstanding content can cut through the noise. Outstanding content can generally be identified because it adheres to the following six rules:

1. IT'S NATIVE

Though the functions of every platform may sometimes overlap, each one cultivates a unique language, culture, sensibility, and style. Some support text-heavy content; others are better suited for richly designed visuals. Some allow hyperlinks; others don't. These differences are not minor—putting the wrong kind of content on a platform will doom your marketing efforts. This should be self-evident, but as you'll see from the examples in this book, many companies just don't take the time to learn the platform's native ways before throwing content on it. Those who do, however, see results. And the ones who really dig deep to understand the subtleties and nuances of the platform that aren't obvious to the more casual user? They truly shine. It's like the difference between someone who learns a new language well enough to order meals in restaurants and talk about their day, and someone who is so fluent he dreams, curses, and makes love in that language. Marketers who understand platforms at that fluent level are the ones whose businesses will be most noticed and appreciated. This has always been the case. People forget that it took a long time for television ads to become as persuasive, and as pervasive, as they are now. Originally, only select families had access to television, and when they did, it was a guy in a suit sitting at a desk heralding the commercials, or a disembodied voice announcing, "This program brought to you by . . ." Not too compelling. Television ads only started to drive sales once TV units made it into more homes and became a popular source of family entertainment. In particular, ads started to work when a few smart marketers figured out how to talk to their consumers in ways that were native to the platform—through short, scene-driven stories populated with evocative characters. The ads became an intrinsic part of the television-watching experience. People hummed the jingles on their way to work or while vacuuming the house. The brands became cultural touchstones, and their products—the Cream of Wheat, the floor wax, and the frozen dinner—flew off the shelves. All because marketers figured out how to create content that was visually compelling, story-driven, and entertaining—ads that mirrored the content already airing on the platform and that the television audience was coming to see.

Content is king, but context is God. You can put out good content, but if it ignores the context of the platform on which it appears, it can still fall flat. Most marketers are oblivious to context because marketers are on social media to sell stuff. Consumers, however, are not. They are there for value. That value can take many forms. Sometimes it's in a few minutes' respite from the stress of a busy day. Sometimes it's in the form of entertainment, information, news, celebrity gossip, friendship, a sense of connection, a chance to feel popular, or an opportunity to brag. Social networking sites light up people's dopamine pathways and the pleasure centers of their brain. Your content must do the same, and it will if it looks the same, sounds the same, and provides the same value and emotional benefits people are seeking when they come to the platform in the first place. In other words, it will if it is native.

What is native to a platform? Depends on the platform. Tumblr attracts the artsy crowd and supports animated GIFs (short, rolling loops of video). A text post from a design firm reading "Visit our Web page to see our award-winning office furniture designs," would be wasted there (actually, that would be a lousy post on any platform). So would a low-quality photograph on glossy, picture-perfect Pinterest. Twitter speaks to an ironic, urban audience that loves hashtags. An earnest post like "We love our customers!" would probably be soundly ignored. It sounds funny here, and yet posts like these are everywhere, proving that most brands are ignorant about what is native to a platform.

You already know that successful social media marketing requires throwing many jabs before converting the sale with a right hook. Counterintuitively, the most effective jabs are actually the gentlest. They are thrown with "native" content, which seamlessly blends in with the platform's offerings and tells stories that engage the consumer at an emotional level. From the outside, jabbing with this kind of content won't look or feel like the setup for that selling right hook, but it is, because the long-term financial worth of a person's smile, giggle, snort, and even her tears is invaluable.

Native content has been compared to a modern-day version of advertorials, or infomercials. Just like the talk show that isn't really a talk show, but a venue for selling slow-cookers, or the headlined article that isn't really an article, but an introduction to a new joint-pain medication, native content looks and sounds exactly like any other content that appears on the platform for which it was created. The similarities, however, stop there.

Infomercials and advertorials are usually ridiculed because of their poor production value. There's something cheesy about them. Sometimes that cheese is part of what makes the piece work—it's hard to take your eyes off Ron Popeil puttering around his staged kitchen gabbing with his cohost and pulling chickens out of the Showtime Rotisserie. But classic advertorials and infomercials are hardly subtle—they are loaded with right hooks. They're informative and entertaining, like a jab, but they're there to sell. Whether the brand places its ad on a TV screen or in a magazine, it makes sure to plaster a huge phone number and URL across the bottom. And even if those obvious signs weren't there, the whole tone of the piece is that of a sales pitch. Consumers couldn't avoid the sell if they tried.

Native content, however, is not cheesy when it's done right, nor is it obvious. What it is, really, is cool. Now, what is the formula for cool? Beats me. You know it when you see it. It's whatever hits your emotional center so hard you have to share it with someone else. It can be a quote, a picture, an idea, an article, a comic strip, a song, a spoof, but whatever it is, it says as much about you, the person sharing it, as it does about the brand or business that originated it. There is no formula for cool content, other than that you can't make it if you don't have a deep understanding of what makes your audience tick and what they're seeking when they use social media.

Creating skillful native content has little to do with selling and a lot to do with skillful storytelling. In the right social-media-savvy hands, a brand that masters native content becomes human. Though of course the topics of Campbell Soup Company's posts on Facebook will probably be vastly different from your mother's, they should still look and feel like something a real person, whether a friend, acquaintance, or expert, would write. When native content is skillfully delivered, a person will consume it with the same interest as he would anyone else's. That's because unlike most of the marketing tactics forced down consumers' throats in the past, smart, native social media tries to enhance the consumer's interaction with a platform, not distract him from it.

You see the difference? For more examples, check out the color commentary at the end of chapters 3 through 7.

NATIVE	NON-NATIVE
Bud Light Facebook	**Best Buy Facebook**

Sesame Street Tumblr	**Sharpie Tumblr**

2. IT DOESN'T INTERRUPT

The Keebler Elves, the Trix bunny, the Yoplait ladies one-upping each other with ecstatic proclamations of how good the yogurt is—they were all created to entertain, so that the next time you were in the mood for cereal or a snack, you'd remember the funny ad and be compelled to try the product. The Marlboro Man's steely jaw and far-off stare were designed to convince you that if you smoked his cigarette, you too might exude an ounce of his masculine, independent essence. Ads and marketing are supposed to make consumers feel something and then act on that feeling. In that regard, the content marketers create today is similar to what it would have been fifty years ago. Where it should differ, however, is in the way it affects, or rather, doesn't affect, your consumer's media experience. Despite being the strong, silent type, the Marlboro Man was still an intruder. People would be watching *Bonanza* and then there he'd be, interrupting their program to sell them cigarettes. Then ads for Pine-Sol, Bengay, or Jif would follow. No matter how good the ads were, there was a distinct break between the show people were watching and the ad. But today marketers don't have to intrude on the consumer's entertainment. In fact, it's imperative that we don't. People have no patience for it anymore, as evidenced by the speed with which they jumped on the chance to bypass advertising altogether with the advent of DVRs in the late 1990s, and other commercial-skipping devices. If we want to talk to people while they consume their entertainment, we have to actually *be* their entertainment, melding seamlessly into the entertainment experience. Or the news experience. Or the friends-and-family experience. Or the design experience. Or the networking experience. Whatever experience people are seeking on their preferred platforms, that's what marketers should attempt to replicate. They may not be in a buying frame of mind today, but you never know about tomorrow, and they will be far more likely to make a purchase from a brand they believe understands them and represents what they value than one to which they have no emotional connection.

3. IT DOESN'T MAKE DEMANDS—OFTEN

Advertising impresario Leo Burnett offered the following advice for making great content:

Make it simple.
Make it memorable.
Make it inviting to look at.
Make it fun to read.

I'm going to add one more directive: Make it for your customer or your audience, not for yourself.

Be generous. Be informative. Be funny. Be inspiring. Be all the characteristics we enjoy in other human beings. That's what jabs are all about. Right hooks represent what is valuable to you—getting the sale, getting people in the door. Jabs are about what is valuable to the consumer. How do you know what content people find valuable? Look on their phones. Phone home screens show you everything you need to know about what kind of content people value. In general, the three most popular app categories are:

a. Social networks, which tells you that people are interested in other people.
b. Entertainment, including games and music apps, which tells you that people want to escape.
c. Utility, including maps, notepads, organizers, and weight loss management systems, which tells you that people value service.

Much of your content should fall within one of these three categories. Sometimes the possible jabs a business should take with this content will be obvious. A cosmetics company could easily tell a story about utility by giving their customers short videos (under fifteen seconds) on Facebook on how to properly apply their makeup, or put out an infographic on Pinterest illustrating the interesting facts about their product history and how women have used it over time. But how would a cosmetics company provide entertainment? If it's selling to eighteen- to twenty-five-year-old females, it could post demos of new music that appeals to eighteen- to twenty-five-year-olds, and deconstruct female music stars' stage makeup, maybe admiring the risks they take and explaining how people could try to get the same toned-down effect at home. As for how the company can tap into its customers' desire to interact with people, it just needs to be human. It needs to get in on conversations, find shared interests with consumers, and respond and react to what people are saying, not just about the brand per se, but about related topics, like how women can erase the signs of fatigue and stress before a big presentation even when they've been up since three in the morning with a baby, or what age is appropriate for girls to start shaping their eyebrows. It could also talk about unrelated topics. Just because its main product was makeup wouldn't mean that it couldn't also talk about gaming or food, because it's possible that fans could be enthusiastic about those topics, too. Jabs can be anything that helps set up your "commercial ask."

When you deliver a precise jab with native content, it might take your consumer a split second before he realizes that the story he's paying attention to is being told by a brand, not an individual. Yet if your content is great, the realization won't piss him off. Instead, he'll appreciate what you're offering. Because when you jab, you're not selling anything. You're not asking your consumer for a commitment. You're just sharing a moment together. Something funny, ridiculous, clever, dramatic, informative, or heartwarming. Maybe something featuring cats. Something, anything, except a sales pitch. Skillful, native storytelling increases the likelihood that a person will share your content with a friend, thus increasing the likelihood of that friend remembering your brand the next time she decides she needs whatever it is you sell. It might even increase the chance that when you finally do hit her with a right hook and ask her to buy something from you, she will click through to make an immediate purchase, even though she's sitting under a dryer at the salon (this moment brought to you thanks to the generous contribution of mobile device developers everywhere).

The emotional connection you build through jabbing pays off on the day you decide to throw the right hook. Remember when you were a kid, and you'd go to your mom and ask her to take you out for an ice-cream cone, or to the video arcade? Nine times out of ten, she said no. But then, every now and then, out of the blue, she would say yes. Why? In the days or weeks prior, something about how you interacted with your mother before the unexpected outing to the ice-cream shop or arcade made your mom feel like she wanted to do something for you. You made her happy, or maybe even proud, by giving her something she valued, whether it was doing extra chores or good grades or just one day of peace with your sibling. You gave so much that when you finally asked, she was emotionally primed to say yes.

No way is a consumer going to say yes if

you ambush him with a giant pop-up that blacks out the middle of the Web page he's reading. The only thing he'll feel is irritation as he frantically hunts for that little X in the corner that will make you go away. If consumers could wipe out all the banner ads blinking around the periphery of their Web pages, too, they would. No one wants to be interrupted, and no one wants to be sold to. Your story needs to move people's spirits and build their goodwill, so that when you finally do ask them to buy from you, they feel like you've given them so much it would be almost rude to refuse.

Jab, jab, jab, jab, jab . . . right hook!

Or . . .

Give, give, give, give, give . . . ask.

Get it?

4. IT LEVERAGES POP CULTURE

There's a great scene in the movie *This Is Forty* where two parents tell their daughters they're going to eliminate the Wi-Fi so the family can bond better without the distraction of electronics. For entertainment, the mom and dad suggest building a fort, or running around in the woods, or putting up a lemonade stand. The girls have no idea what their parents are talking about; without their phones, they may as well be condemned to life in an isolation cell. Histrionics ensue.

It's no joke. Generations are defined by their pop culture, and without it, they're lost. Take away a young person's tech and you've taken away her lifeline to everything that matters to her. In days past, kids met their friends at the soda fountain and listened to records. Then they hung out at the mall and listened to cassettes. Later they hung out at the 7-Eleven parking lot and listened to CDs. Now they hang out on their phones, simultaneously listening to downloads, checking the celebrity news, chatting with their friends, playing games, all on their smartphones and tablets. And your content has to compete with all of it. But as the saying goes, if you can't beat 'em, join 'em. The young generation isn't the only one consuming their culture via phone, either. Everyone is, including the ones who used to listen to their music on records, cassettes, and CDs. So use that to your advantage. Show your fans, whoever they are, that you love the same music they do. Prove that you understand them by staying on top of the gossip about celebrities from their generation. Create content that reveals your understanding of the issues and news that matter to them. Just don't place it in a mobile banner ad. The days of stopping people from what they're doing to look at your ad are at best diminishing, and more than likely over, and regardless are overpriced for the ROI. Integrate your content into the stream, where people can consume it along with all their other pop culture candy.

5. IT'S MICRO

There is something else you could do as you reevaluate your social media creative: stop thinking about your content as content. Think about it, rather, as micro-content—tiny, unique nuggets of information, humor, commentary, or inspiration that you reimagine every day, even every hour, as you respond to today's culture, conversations, and current events in real time in a platform's native language and format.

A well-known (in advertising circles) yet perfect example of micro-content practically stole the show at the 2013 Super Bowl. When the power went out in the Superdome during the third quarter, leaving thousands of spectators in the dark for a half hour, while the players for the Baltimore Ravens and San Francisco 49ers hunkered down trying to keep their bodies limber and their heads in the game, Oreo saw an opportunity. It tweeted, "Power Out? No Problem." Attached was a photo of a lone Oreo cookie waiting in the dark, with accompanying text that read, "You can still dunk in the dark." Suddenly, all those people in limbo waiting for the power to be restored and the game to start over saw a funny reminder that Oreo is the cookie for all occasions. The tweet didn't tell anyone to go buy Oreos. It didn't include any call to action, actually. It didn't need to. Within minutes it had been retweeted across Twitter and liked on Facebook tens of thousands of times. Why? No one had ever seen anything like it. It's one thing for a Ravens fan or a 49ers fan to tweet or post status updates chronicling her reaction to the game; we've gotten used to seeing individuals respond to real-time events around the world. But to see a brand do it as casually and naturally as a real person? That was a first for such a mass-market brand within the context of such a mainstream event. The tweet was only possible because Oreo had thought far enough ahead to have a social media team at the ready to respond to whatever happened on television. Talk about proper investment in a platform. Key to the ad's success was not only the fact that it was clever and elegant, but also that it aligned perfectly with Oreo's brand identity, as well as the identity of Oreo lovers everywhere. Oreo is the playful cookie, the fun cookie, the cookie you want to watch football with.

Did the micro-content offer consumers anything of value, as a proper jab should? It's unlikely it would have received any attention if it hadn't. Don't underestimate the value of a fun surprise, a grin, and a sudden craving for chocolate and shortening. For a few days, the whole world, in traditional media and social, had positive things to say about Oreo. At the very least, everyone who saw it got the chance to say they witnessed the beginning of a new era in marketing.

The next time a brand responds in real time, will the Twittersphere go crazy? Probably not, which is why it's good to be first to market, even on platforms that don't appear to have tremendous value at first glance. Your job as a marketer is not just about selling more product (though that's a priority, and don't forget it), but increasingly about making sure that you are first to market as often as possible in terms of timing, the quality of your micro-content, and the originality with which you respond to the world around you. This is true no matter what platform you work with, from Twitter to Facebook, from Instagram to Pinterest.

Oreo's strategy through the Super Bowl exemplifies the only formula for social media success that doesn't change depending on platform or audience:

Micro-Content + Community Management = Effective Social Media Marketing

Some people weren't impressed by the tweet. Imagine, using a platform the way it's supposed to be used! But so few companies manage it, it is worth applauding when one succeeds. This move took a lot of forethought. Oreo had to have a team in place, watching, waiting for the first opportunity to strike as the game went on. Old Spice managed something similar a few years ago with its "The Man

Your Man Could Smell Like" campaign, in which the actor Isaiah Mustafa replied to consumer questions in real time on the Web. But that Q&A was the result of a carefully orchestrated campaign. Oreo had a TV ad that ran during the Super Bowl (and integrated Instagram) but otherwise had no plan other than to be in a position to respond to real-time events in real time. That's hard to do, and they did it perfectly, keeping things simple, immediate, and relevant.*

Businesses can forge a direct connection between their community and their brand when they stop thinking about social media as the backup to the main events. It should be a main event in and of itself, serving as the nexus connecting every other channel by which businesses talk to their customers.

There's no reason for marketers to draft new overarching social media campaigns every year. Everyone's should be as simple as this:

Jab at people, all the time, every day.

Talk about what they're talking about.

When they start talking about something different, talk about that instead.

Repeat.

Repeat.

Repeat.

Not every brand has to jab at the same rate as its competitor. Remember, quality and quantity—some brands can get away with just a few jabs here and there; others need to jab all the time. I don't have to jab nearly as often as I did when I was first starting out. BP doesn't have to jab as often as it did after the Deepwater Horizon oil spill in 2010. Apple probably didn't have to jab at all at the height of the iPhone frenzy, when the product was still new. Successful storytelling builds brand equity, and businesses with high brand equity don't need to draw as much attention to themselves and their achievements as those that are still establishing their value to the consumer. Yet even if you don't have to jab frequently, you can't ever stop entirely, and you certainly can't stop watching for those special opportunities where your brand can take advantage of breaking news or the culture at large to prove its relevance or show it's paying attention. Social marketing is now a 24-7 job.

---

* Go to JJJRH.com/oreotalk for a free one-hour video of the team behind the campaign and me speaking at a SXSW 2013 panel.

6. IT'S CONSISTENT AND SELF-AWARE

Consider how each and every post, tweet, comment, like, or share will confirm your business's identity. Though your business's micro-content will vary wildly every day, it must consistently answer the question "Who are we?" You can and should learn to speak as many languages as possible, but no matter which language you're using, your core story must remain constant. And no matter how you tell your story, your personality and brand identity must remain constant, too.

When you're self-aware, you know your message. When you know your message, it's easy to keep it consistent in every setting. No marketer should find this a daunting concept—we do it every day when we navigate the analog world. You're going to wear a different outfit and use different vocabulary when you're sitting down for tea with your grandmother in her home than when you're living it up with friends in a nightclub. At least, you will if you've got nice manners. Creating micro-content is simply a way for your brand to adapt according to the circumstances and the whims of your audience. Micro-content is your brand's best chance of being noticed in an increasingly busy, disjointed, ADD world.

When you create stellar content native to a platform's context, you can make a person feel; if your content can make a person feel, he is likely to share it with others, providing you with amplified word of mouth at a fraction of the cost of most other media. Best of all, you not only own the content, you own the relationship with your customer. You're not spending a million dollars to rent it for thirty seconds from a television network. You could spend a million dollars to acquire committed fans on Facebook, and that would be money well spent, but if you also storytell properly, the only additional cost you'll have is for the nonworking creative. Your content simply lives on, replicating itself over and over as your fans and followers pass it along through word of mouth, diminishing your costs with every retweet, share, pin, heart, and post. The concept of owning content and relationships instead of renting them has gained enormous traction with the start-up entrepreneurs of Silicon Valley, but it has been slower to infiltrate the mindset of most Fortune 500 companies and traditional small businesses around the world. That's going to change once they realize that they are no longer beholden to media companies to disseminate their content and connect with their consumers. Thanks to social media, they'll be able to do it all by themselves. Some already are, as we'll see in the upcoming chapters.

ROUND 3:
STORYTELL on FACEBOOK

- Founded: February 2004
- The platform was called Thefacebook.com until August 2005.
- In a 2006 survey of the top five "in" things on college campuses, Facebook tied with beer but scored lower than iPods.
- The "Like" button was originally supposed to be called the "Awesome" button.
- Mark Zuckerberg initially rejected photo sharing; he had to be persuaded that it was a good idea by then-president Sean Parker.
- There were more than a billion monthly active users as of December 2012.
- There were 680 million monthly active users of Facebook mobile products as of December 2012.
- One out of every five page views in the United States is on Facebook.
- Let me say that again: ONE OUT OF EVERY FIVE PAGE VIEWS IN THE UNITED STATES IS ON FACEBOOK!

What more could possibly be said about Facebook? We all know what it is and what it does. We all know it's the biggest, baddest social network, the one that changed our culture as monumentally as television. While

still skeptical about most other social media platforms, small business owners, marketers, and brand managers consider Facebook a legitimate marketing tool, though, strangely enough, not because it has the most sophisticated analytics available. Rather, they trust it because it's hard to dismiss a platform as skewing too young, or too experimental, or too trendy, when your niece, your brother, your seventy-two-year-old dad, and more than a billion other people are on it. Familiarity breeds acceptance. Only the most stubborn holdouts, mostly from companies working B2B or just contrarians, question whether their customer is actually on Facebook and whether it's worth maintaining a presence there.

It stands to reason that if this is the platform with which most people are familiar, it's the one that requires the least explanation. Yet this chapter ended up being the longest in this book, because although most marketers think they understand Facebook, they obviously don't. If they did, consumers would be seeing much different content, not just on Facebook, but across all platforms. For now, however, the majority of brands and businesses still haven't realized the unprecedented insight Facebook gives us into people's lives and psychology, insight that allows marketers to optimize every jab, every piece of micro-content, and every right hook.

Think about why people go to Facebook: to connect, socialize, and catch up on what the people they know and presumably care about are doing. In the process, they also find out what their friends and acquaintances are reading, listening to, wearing, and eating; what causes they are championing; what ideas they're hatching; what jobs they're hunting; and where they are going. Facebook wants users to see things that they find relevant, fun, and useful, not annoying and pointless, or else they'll abandon the site. Which means you'd better create content that's relevant, fun, and useful, too.

Now, if it were that easy, this really would be a short chapter. Hire better creatives, make better content, and you'd be good to go. The problem is that there are three forces that have made it more difficult than it used to be for even the most talented creatives to organically deliver awesome content on Facebook: the masses, the evolution of the masses, and Facebook's response to the evolution of the masses.

The very thing that makes marketers want to have a presence on Facebook—the sheer number of users—makes the platform a marketing challenge. A billion users, and all the content they generate, creates a conundrum: with so many pieces of content streaming into consumers' News Feeds and competing for attention, it's unlikely they will see any content you post, even the good stuff.

In addition, users are human. They age and mature. They grow up, break up, have kids, quit the guitar, take up fencing, or go vegetarian. The user who became your fan in 2010 will not be the same fan in 2014. But even though he's changed, he probably hasn't thought to go back and remove outdated information about his tastes and preferences on Facebook. We're always going to follow more people and brands than we need to. We may not be watching this TV show anymore nor following that actor, but we don't unfollow their pages as we move on in life. As those bygone interests fade from our consciousness, we expect them to fade from our pages and News Feeds, too.

Facebook knows this. Long ago, when college students were the biggest population on Facebook and the user pool was relatively small, people's News Feeds were organized chronologically. But as the user base grew—and grew and grew—Facebook had to figure out how to prevent users' streams from getting clogged up with posts they weren't interested in. It didn't want to be Twitter, with its waterfall of content from every person, organization, brand, and business in which users ever expressed interest; it wanted to curate our News Feed and make sure the majority of what we saw was always important and relevant to us. To help mitigate the consequences of literal TMI, Facebook finally settled on an algorithm called EdgeRank. Every interaction a person has with Facebook, from posting a status update or a photo, to liking, sharing, or commenting, is called an "edge," and theoretically, every edge channels into the news stream. But not everyone who could see these edges actually does, because EdgeRank is constantly reading algorithmic tea leaves to determine which edges are most interesting to the most number of people. It tracks all the engagement a user's own content receives, as well as the engagement a user has with other people's or brands' content. The more engagement a user has with a piece of content, the stronger EdgeRank believes that user's interest will be in similar content, and it filters that person's news stream accordingly (a randomizer ensures that occasionally we'll see a post from someone we haven't talked to in years, thus keeping Facebook fresh and surprising). For example, EdgeRank makes sure that a user who often likes or comments on a friend's photos, but who ignores that friend's plain-text status updates, will see more of that friend's photos and fewer of his status updates. Every engagement, whether between friends or between users and brands, strengthens their connection and the likelihood that EdgeRank will push appropriate content from those friends and brands to the top of a user's News Feed. That's of course where you, the marketer, want to see your brand or business.

That's why it's never been more important to produce quality content that people want to actually interact with—a brand's future visibility on the platform depends on its current customer engagement levels (and soon this trend will spread to all the other platforms, as well). Unfortunately, the engagement that marketers most want to see—purchases—is not the engagement that Facebook's algorithm measures, and therefore not the engagement that ultimately affects visibility. More than anything else, marketers want users to respond to their right hooks. That's why they put so many out there. What they don't realize, however, is that on Facebook, it's the user's response to a jab that matters most.

Here's why: Through EdgeRank, Facebook weighs likes, comments, and shares, but it currently does not give greater weight to click-throughs or any other action that leads to sales. EdgeRank doesn't care, actually, whether you sell anything, ever. Facebook's greatest priority is making the platform valuable to the consumer, not to you, the marketer. What it cares about is whether people are interested in the content they see on Facebook, because if they're interested, they'll come back. What proves interest? Likes, comments, shares, and clicks—not purchases. You could put out a piece of content with a hyperlink to your product page that garners $2 million in sales in thirty minutes. Facebook would take note of the heightened interest, and the algorithm would push you to the forefront of your current fans' News Feeds. But link clicks do not create stories, so if no one shares that piece of content, or even likes or comments on it, the content will reach your current community, but Facebook will not deem it interesting enough to show it to a wide number of people outside that. If you want to maximize your eyeballs, it's not enough to get people to read your article or buy your product—you have to get them to engage with it so that it spreads. On Facebook, the definition of great content is not the content that makes the most sales, but the content that people most want to share with others.

Unfortunately for marketers, as with all platforms that you can't test in a controlled environment, it is still difficult to make a direct correlation between high levels of engagement and sales. However, it stands to reason that the only way you can make any sales is if as many consumers as possible see your content (and if customers are seeing it, it had better be what you want them to see). Consumers' eyes are on Facebook. If the only way to reach those consumers is to get them to engage, then it's up to you to create not just great content, but content that's so great they want to engage with it. To put it in boxing terms, you have to jab enough times to build huge visibility, so that the day you do throw a right hook—the day you do try to make a sale, say, with a post that's not particularly shareable but where the link takes people to your product—it will show up in the maximum number of News Feeds.

Unfortunately, while it tries hard to guess what is important to users, Facebook still can't determine their intent. Which action, or edge, indicates more interest—commenting on a post or liking a post? If a person actually clicks on a picture, is she showing more interest than if she shares it? Is a picture more valuable than a video? Does liking a video post show equal interest as watching the entire video? Facebook doesn't know, but it desperately wants to, so it keeps tweaking the algorithm to figure the mystery out. This is why even though most of your content might get seen today, you can't trust that it will tomorrow. One minute your brand could be popping up at the top of a user's page; the next it could be buried six pages down. For example, Facebook may decide that sharing is a much stronger call to action and brand endorsement than liking, so it will give sharing more weight than a like. If your content happens to elicit many shares, you're golden. But then Facebook could change its mind and decide that likes are actually as valuable if not more so than shares. Your content doesn't usually get that many likes. Now what?

The speed with which we have to keep up with these changes, and create matching content, is enough to give even the most seasoned marketer a case of whiplash. How are we supposed to jump through the hoops to reach our consumers if Facebook keeps moving the hoops around?

By staying vigilant. By accepting that you're going to reinvent your content every day, if not more. And by getting to know your community like your own family. How do you do that? You tell them stories they want to hear. You give openly and generously. You jab, jab, jab, jab, jab.

JABS IN ACTION

The key to great marketing is remembering that even though you're all about your brand, your customer is not. As with any first date, getting a second date depends on you doing your best to learn more about what the other person is interested in, and directing the conversation in that direction. In the end, boxing and dating are really not that different. After all, the goal is to score. Sometimes the score is measured in points, and sometimes in a marriage proposal (or something else), but in either case you won't win if you play your most aggressive move first.

Let's say your company sells boots. It would make a lot of sense for you to talk about weather. It would make a lot of sense to talk about rock climbing. It would even make sense to talk about hunting or maybe even something like how the boots protect people's feet during rowdy concerts. These are all topics that are directly related to boots, or at least only about one mental step away. So for your first jab, you put out the following status update:

"So long, *30 Rock*! Thanks for seven hilarious years!"

If the CMO of this boot company knows only as much about social media as the average businessperson, as soon as she sees that first status update she's going to storm up to you and question the living crap out of it. What does *30 Rock* have to do with our boot company? How off-brand can you get? Why are we doing this? How does this sell more boots? And your answer will be, it doesn't. Yet.

As the CMO of the boot company stands there looking, at best, curious and, at worst, furious, you will calmly point to the analytics (called Page Insights), which will reveal that that particular post is getting higher than usual engagement over more traditional boot-centered posts, just as you thought it would. Why? Because through previous jabs asking things like "What's your favorite TV show?" you had already gathered the consumer insight that 80 percent of your fans were crazy about *30 Rock*. And you knew that the series finale was approaching. So by putting out a "Good-bye, *30 Rock*" piece, you are connecting with your community and showing them that not only do you get them, but you are one of them. All of a sudden your brand is talking like a human being, not a boot company. And as the overindexing (meaning a post performs above normal for that brand) reveals, people like that. They respond. This is good for you, because the uptick in engagement tells Facebook that this brand matters to people. So when you put out your next piece of content, a fifteen-second user-generated video of people showing off their boots, Facebook makes sure your customers see it in their News Feed. Again, the piece isn't selling anything. Nor is the next one, a Valentine's Day card that doesn't show a single boot. Then you put out another three or four pieces of content that don't sell anything, either, like this:

Third jab: Post—A fifteen-second video about rock climbing.

Fourth jab: Poll—"Would you rather wear your boots in the summer or the winter?"

The point is to give and give and give, for no other reason than to entertain your customers and make them feel like you get them. And the more you give, the more you really will get them. Before, every piece of content had to be a right hook because all we knew about customers who bought boots was that they needed protective footwear. But if we jab wisely, Facebook can give a detailed and nuanced understanding of the people who buy our products. By testing and jabbing and giving, we learn what they find entertaining. Content that entertains sees engagement. Content that sees engagement tells Facebook and the rest of the world that your customers care about your brand, so that when you finally do put out something that would directly benefit your bottom line—a coupon, a free-shipping offer, or some other call to action—4 percent of your community sees it instead of a half percent, which gives you a much better chance at making a sale.

TARGET YOUR JABS AND RIGHT HOOKS

Sometimes, though, you don't want everyone to see the same information. On any other platform, where your posts are entirely public, every jab hits everyone in the face. On Facebook, however, you can be extremely selective, customizing your jabs and targeting subsets of your fan base. Want to target a post for thirty-two- to forty-five-year-old married women with college degrees who speak French and live in California, and post it on New Year's Eve? When you know how to use Facebook properly, you can (and I imagine the largest liquor store in California would).

Targeting your posts is a strategy to keep in mind when you're jabbing; it's flat-out essential when you're throwing a right hook. Let's say you're a national fashion retailer, and today is Black Friday. You've created a piece that highlights one of your most coveted purses. You know that the buyers of that purse are generally twenty-five-year-old females. Does it make any sense to send that content about a purse to your fifty-five-year-old male customers who primarily come to you for belts? Of course not. So when you post the announcement about tonight's Black Friday sale, with a picture of the purse, you post it only to fans of your page who are twenty-five- to thirty-five-year-old women. By speaking directly to the right demographic, you've increased the probability that people will engage with that content, which keeps your EdgeRank numbers up, instead of giving Facebook the impression that people don't care about your brand anymore by posting it to men who are never going to click or engage with a post about a purse.

Now, you could post the piece to your fifty-five-year-old male customers if you change the content so that it resonates with them. Maybe it reads, "Hey Dad, it's never too late to remind her that she's still your best girl. Our Black Friday sale starts tonight, 6:00 P.M." You go even further and design the content so that it goes out to consumers in Texas in the shape of Texas, and the content that goes to New Jersey is in the shape of New Jersey, and so on and so forth for any of the states whose residents have a particularly strong streak of state pride. For any jab or right hook to have impact, it has to speak to the consumer and hit his or her emotional center.

SMART SPENDING

It's worth taking a step back and examining the cost-effectiveness of this scenario. With very little lead time, a retailer can create two distinct pieces of content, send it directly to two separate demographics, and watch in real time to see how the recipients respond. If the excited comments start to pile up, or the content starts getting shared, that retailer knows the right hook made its mark. Its consumers engage, thus kicking up the retailer's EdgeRank, which shows Facebook that its users value the retailer. It makes sure the content shows up in more people's stream, which therefore allows the retailer to show its content over and over again to an ever-larger audience without having to pay any more for it.

To accomplish the same thing on television, a national retailer might create two different TV spots targeting different demographics. For example, it would launch one mainstream targeted ad that would run on CNN during primetime, and a multiculturally targeted ad that would run on UPN channels during the local 10 P.M. news. The creative team would have to develop the ads weeks before they ran. Typically, the spot would need to run enough times so that the retailer's desired reach population would have seen the spot three times—about a two-week flight of spots. It would cost the retailer between $7,000 and $13,000 to reach this audience. Then, once the pieces had run, it would have to sit and cross its fingers that people had actually watched the ad even though they had just forgotten to turn the TV off while streaming a movie on their second screen. And if it wanted to run more content, it would have to pay all over again.

Which scenario sounds more time- and cost-efficient to you?

Now, there's nothing wrong with spending money when you're spending it smartly. All along you've probably been buying the Facebook ads that line up along the right side of the site. Those ads have until now been one of the most efficient ways to spend dollars for any brand or business, big or small. On average, the cost of running an ad on the right side of the page on Facebook runs the gamut between $.50 to $1.50 per like, though depending on the specificity of your targeting, the length of your campaign, and your budget it's possible to acquire likes for as low as $.10 and as high as several dollars. That's a steal, even when you compare it to the cost of email acquisition, which can run as low as $0.49. How can a dollar spent acquiring a Facebook fan be worth more than forty-nine cents anywhere else? Because a social user on your fan page has more potential reach than anywhere else.

I should know. Back in 1998, I was using email marketing, as well as search engine marketing (SEM) and pay-per-click ads, to build WineLibrary.com. People loved my product and my business and were happy to subscribe to my emails and to buy from me. My business model then was no different from that of any of the successful email marketing companies of the last half decade like Fab.com, Groupon, or Gilt. The difference is that their fans aren't as beholden to their email as mine were in 1998. If my fans wanted to talk to or share information with friends, they had to use email. Today's fans don't. So today's email marketers have had to offer huge rewards for sharing, such as $10 off a first order if the customer can get five friends to subscribe to the site. Without that incentive, people won't spread content or invite friends to join them on your site via email—it feels too much like spreading spam. Social media, however, is built for sharing, so those targeted Facebook ads, though costing $.50 to $1.50 per fan, are actually worth much more because those fans are more likely inspired to share your content for free, and possibly more than once—if you give them what they want in terms of content and service.

THE CHANGING FACE OF SMART SPENDING

Unfortunately, Facebook ads in their current incarnation are going the way of the dinosaur, and the days of cheap fan acquisitions are coming to an end. With the substantial growth of mobile for Facebook and the increase in people abandoning their laptops, the ads on the right side of Facebook's desktop are becoming obsolete. You could hope that consumers will think to go directly to your fan page for a steady stream of your content, but honestly, unless you're doing research, do you go to that many fan pages just for kicks? Probably not. And we're all going to do it even less now that we're spending more time on Facebook's mobile app than we do on the website itself.

There is no substitute for the real estate of a desktop on a mobile device—there's no room. This means that until the next great technological revolution, like Google glasses or tattooed screens in the palms of our hands, all of your Facebook stories, content, and marketing must be developed for the mobile experience. This is why in January 2013, Facebook CEO Mark Zuckerberg announced that Facebook should now be considered a mobile company.* And just six months later, Facebook reported 41 percent of its ad revenue came from mobile, equaling $1.6 billion in the second quarter of 2013.

But if marketers are limited to smartphone screens, where are marketers supposed to put their ads? Some brands have decided that the answer is: right on top of the page the consumer is trying to read. It

* It's a challenge and an opportunity when you are writing books as current events happen. For the record, that talk was given exactly two days before this page was first drafted.

has surely happened to you—you head to your favorite site to check the news, and instead of seeing your content, a big intrusive box overtakes the screen, pitching electronics or software or something that you did not come to the site to see. Why do marketers think this is a great way to get people to do business with them? All is does is piss people off and elicit negative feelings toward your brand. It is the antithesis of jabbing. All impressions are not good impressions. Quality, relevance, good timing—these things matter far more than many marketers realize. Once again, we have to keep in mind why people gravitate to Facebook, or any site, for that matter. It's not to see ads.

So what's a marketer to do? We need to rethink what an ad looks like, and what it accomplishes. We need to go native. We need to bring value. From now on, the difference between your content and your ads on Facebook will be . . . nothing. Your content, or rather, your micro-content, has to *be* the ad. Fortunately, Facebook has been perfecting a tool that allows you to create ads out of content that has already been vetted by your fans, which will not only help you improve your content's reach, but will actually protect you from putting out content that is simply a waste of your and your customer's time. It's called a sponsored story. And unlike a TV ad or magazine spread, this spending strategy is worth every penny.

SPONSORED STORIES

Sponsored stories were launched in early 2011, but it was in the fall of 2012 that they came into their own, mostly because Facebook announced that it was finally making an algorithmic adjustment that would purposely limit how many people would organically see a brand's posts, even if they had already become fans by liking the brand's page. Until recently, though the algorithm was calibrated to limit spam or uninteresting content, good content could still organically reach a large percentage of fans. As of September 2013, however, Facebook's algorithm will only allow your content to reach about 3–5 percent of your fans. To reach more, you have to post some extremely engaging content. Or, you have to pay. In this way, Facebook is able to protect the consumer's experience by raising the barrier to entry to the News Feed.

A lot of the marketing community didn't see it that way. They were livid. How could Facebook force them to pay more to take advantage of its billion users? How disloyal. How conniving. How capitalist.

Did anyone really think that Facebook wasn't going to figure out a way to make more money? Besides, what else was it supposed to do when the right-side-of-the-page real estate for Facebook ads was disappearing faster than curse words in my keynote speeches as people ditched the wide screens of their PCs for their mobile devices? I didn't understand people's fury. Marketers and business owners who would never get mad about paying hundreds of thousands of dollars to a network to get their ad on TV, even when they'd never know whether the ad had gotten anyone's attention, were having coronaries over having to pay for the same kind of distribution. Unlike TV, your content's reach increases only when you've put out content that people actually want to see and think others do, too. The more people who interact with your content, the more you can amplify the word-of-mouth amplification it receives as their actions are shared with other people. Create great content that gets people to engage and Facebook will let you show that content to more and more people. Create content no one cares about and Facebook will make it as difficult as possible for you put more of it out on its site.

Sponsored stories is a superior ad platform because it rewards nimbleness and quick reaction. When it shows us that a piece of content is resonating, we know to spend money on it. It's so clear. When I think of what I could have done with a service like this back when I was in email marketing, I could just cry for the amount of wine that we could have sold. Let's say that on average about 20 percent of the people who received my emails back then actually opened them, and one day I sent out an email that saw a 21 percent open rate. Then I saw that the wine mentioned in that email was suddenly selling extremely well. Clearly something about that email had made it extra valuable to my audience. How much would that knowledge have been worth? I would have happily paid Yahoo, Gmail, and Hotmail a premium to make sure that the next time I sent out that email, as many people as possible saw it, whether it was by working around spam filters or finding a way for the emails to automatically open when people went to their email accounts. A service like that would have been the greatest marketing tool in the world—heck, are you listening, Google?—and it's close to what you can achieve through Facebook sponsored stories.

Facebook is shockingly bad at explaining sponsored stories, so let me try here.

There are two types. One simply extends your chosen piece of content to the news streams of a larger number of your fans than the regular 3–5 percent that would normally see it. That's called a Page Post. The other extends your reach the same way, but it allows you to highlight the fact that a fan has engaged with your content and tell that fan's friends about it. You can choose to create this kind of sponsored story around a check-in, a like, and several other actions such as when someone shares a story from your app or your website. For example, if a fan checked in to a hotel, or claimed an offer from a T-shirt company, the hotel or T-shirt company could pay to make sure that friends of that fan knew about it, not with an ad lingering on the periphery of the Facebook page that no one but PC users will see, but within the actual newsfeed. That's the big breakthrough for marketers. Before, when we created ads around a post, as soon as it migrated to the right side of the page the format of the post would change. This transformation compromised the impact of the creative work because it no longer looked like an organic piece of content created by someone you knew, but like an ad created by some stranger. But now marketers can keep the creative that we already know works organically, and enhance its power simply by paying to have more people see it, offering us an unparalleled opportunity to connect with active fans as well as reinvigorate relationships with fans that might have gone dormant over time.

Sponsored stories work like this: When I sponsor the story, a higher number of people than normally follow my page will see it in their News Feed. Now they are reminded about me. If the content is actually good enough to compel them to act on that piece of content—liking, sharing, or commenting on it—they get brought back into my orbit, and Facebook believes I am relevant once again: "Facebook users like GaryVee, so I'm going to give them more GaryVee." Now the next time I post a new piece of content, many more of those people will be likely to see it. Yet I won't have had to pay any additional money to get those impressions. And if the engagement continues, my initial costs will continue to diminish as my impressions rise. It could trigger a snowball effect that could last well into the next month, and all for the price of one small sponsored story.

It's important to realize that when you sponsor a story, you don't buy additional data. What you get is extended reach and an additional layer of targeting above and beyond that of an ordinary post or targeted post, both of which are free. Put money behind a well-performing targeted post and turn it into a sponsored story, and you'll increase the specificity with which you can target your audience. You could target a post for women, but your sponsored story can target women who enjoy arts and crafts, and women who listen to country music. If you find out you've got a large swath of consumers in your base who love dubstep, you might want to reference Skrillex in your content and send it their way. If you've created a piece of content with a hip-hop theme, you can check to see which of your fans consistently listens to A$AP Rocky and other hip-hop artists, and only send your content to them. Knowing this kind of detail and using it to tailor content to match your fans' tastes allows you to create pulverizing right hooks.

GREAT BANG FOR YOUR BUCK

The sponsored story is one of the great ad opportunities of all time because it won't let you spend more than your content is worth. Facebook calculates the initial value of your sponsored story based on the competition you face for your targeted audience, and how much that competition is willing to pay. From there, you then tell Facebook how much you're willing to pay for each click or impression you want. But you won't necessarily pay that amount. If you create a great ad that compels people to engage with you, Facebook will decide that your ad deserves priority over a competitor's ad that isn't as engaging. Facebook will let you buy your impressions for cheaper than your competitors if it sees that your ads are performing well, that people like them and are acting on them. In addition, when Facebook sees that people are interacting with your content, it will show that content to more people, because it is obviously enhancing the quality and entertainment value of the News Feed. The second people stop clicking, though, Facebook will stop running the ad as a sponsored story. It will still be visible to a core group of people, but it will be allowed to die a natural death, fading into irrelevance. Unless, of course, you insist on throwing more money behind it. But why would you? This time around, the sponsored story will cost you a lot more, and the results will be the same. Essentially, Facebook purposely makes it cost-inefficient to distribute bad creative.

How cool is that? If you make a stupid television commercial, the network is going to run it as many times as you pay it to. No billboard owner is going to look at your art and say, "Dude, I can't take your money. You won't get a dime of business with that." But Facebook will, not because it's nice enough to protect you from yourself, but because it's savvy enough to protect itself from you. It's in Facebook's best interest for you to put out great content. It wants to monetize, but if users start feeling like they're being spammed every time they go to the site, Facebook will suffer.

If networks could show marketers data that proved that every time they showed the consumer a bad commercial, consumers turned off their TV, TV commercials would be better. That's what Facebook, and all social media, can do for us. Ideally, when Facebook informs you that no one is interacting with your sponsored story, that's your cue to stop and rework the piece, or chuck it altogether. Facebook can't tell you why it's not working—you have to use the data it gives you to figure that out for yourself. Social media gives us real-time feedback from the consumer, which forces us to be better marketers, strategists, and service providers.

And it's still ridiculously cheap. Maybe not as cheap as it used to be, but still a hell of a lot cheaper than a TV ad. And find a television network, radio station, newspaper, magazine, or banner ad provider that lets you test-drive your content for free in the form of organic or targeted posts the way Facebook does.

Ultimately, the changes implemented to Facebook ads only changed how much it costs you to work with Facebook, not how you tell your story. If you're a brand that understands how to jab in ways that bring value to your customers—giving them a moment of levity with a cartoon, or a game to play, or any other escapist content, which then primes them to be open to giving you business when you finally ask with a right hook—you'll win. If you're not, you won't. No matter what Facebook does, ultimately, it's the content that matters. You can sponsor crap, and it won't do anything for your sales. But you don't ever have to sponsor crap. Your Facebook community provides you with an automatic crap filter every time you send out your content for free. Your organic reach may only be 3–5 percent or so, but if a large percent of that organic reach is engaging with your content, you know you've got something good. That's the piece you sponsor. If you put out content and it doesn't get any attention, you know you need to rework it or try something new. Facebook gives you a risk-free method to ensure that you only invest in what's going to improve your business.

Things could change in the future. It's possible that the platform will decide to start using actual purchases as indicators of fan interest more than the engagement of comments, likes, or shares. Obviously, making a purchase is a huge indicator that people want to see your content. That could mean that Facebook becomes as much of a right hook platform as it is a jab platform. If that happens, I predict that Facebook will come up with a way to control right hooks as strictly as it does sponsored stories. The last thing Facebook wants to become is a right hook platform, because it will die.

My advice to marketers is to quit complaining and start creating micro-content worth the money it will take you to successfully reach the customers Facebook is now guarding so carefully. Get more entrepreneurial. Figure out how to work the system and get the most bang for your buck. You can afford to be innovative on Facebook in a way that you can't on almost any other platform that exists.

Let's see how. In the following pages, we'll see some examples of perfect Facebook plays, as well as some almost comical misses.

Please note, the critiques of the following case studies are my opinion only, based on years of experience. I cannot claim any knowledge of any business's agenda or original intent. I'm just calling it as I see it.

COLOR COMMENTARY

AIR CANADA: Ruining a Good Idea

Air Canada · 494,738 like this
March 20 at 1:45pm ·

Lucile Garner Grant, our first ever flight attendant, passed away on March 4 at the age of 102. We wish her family our sincere condolences. An adventurer at heart, we're honoured that she chose to spend some of her years with us.
http://aircan.ca/SMJvQL

enRoute | Q&A with Lucile Garner Grant
enroute.aircanada.com

Claim to fame: The first woman to be employed by Trans-Canada Air Lines (TCA), Garner Grant was a flight attendant from 1938 to 1943. She once rode a dogsled from the airport to a radio station in Fort Nelson, B.C., to fetch a weather

When Air Canada's very first flight attendant, who worked for the airline from 1938 to 1943, died at the age of 102, Air Canada paid her tribute by posting her photograph and a link to an interview their in-flight magazine conducted with her about six months before her death. It should have been a successful jab that engaged a large number of their 400,000 fans. Unfortunately, they blew it.

Here's why:

- **It's not visually compelling.**
- **It's burdened with too much text.**
- **It's a link post when it should have been a picture post.**

It would have made all the difference had Air Canada just taken a little extra time to make this post more visually compelling. Most of us would be thrilled to look as good at 102 as Mrs. Lucile Garner Grant does in her head shot. Yet the two big blocks of text surrounding it water down the impact of the photo. It's too much to expect people to read all that when they're scrolling through their mobile devices at warp speed. By uploading the photograph as a picture post instead of a link post, however, and overlaying the lines announcing Mrs. Garner Grant's death onto the picture itself, Air Canada could have emphasized the photo and simultaneously explained why it was relevant. Next to the photo, they should have included nothing but the subhead of the interview (and maybe a mention of the dogsled), along with a link to the article.

Like this:

That's micro-content right there—compact, intriguing, of-the-moment, and native to the platform. The layout is big and eye-catching enough to make a person scrolling through her Facebook News Feed stop and say, "Damn, 102? Their very first flight attendant? What?" and maybe click through to read the whole interview, which really does offer a fascinating glimpse back in time and would be something many people would be compelled to share with friends. Had Air Canada simply made a few small visual and textual adjustments, they would have had more time to honor one of their employees, and also more time to tell a compelling story about their brand.

JEEP: Evoking the Right Emotions

This picture perfectly encapsulates the Jeep brand. Jeep could not have chosen a better model than the pretty young woman in this photo, with her shades, her flying hair, and her huge smile evoking summer, fun, and freedom. What's cool is that she's not a model—she's someone a fan named Megan Bryant photographed and posted on Facebook. The movement and mood of this picture are striking enough to be worth checking out more closely. One look, and you start to wish you had a Jeep, too.

Jeep
"It's a Jeep thing." Symptoms include wind-tousled hair, a perpetual smile, and feelings of euphoria. (Photo courtesy of fan Megan Bryant.) — with Kaitlyn Brooke Latham, Keith E Brown, Laura Rincón, Ritchie Ritz, Rajeev Elavally, Yahia Mirmotahari, Anay Tobon, Richard Antonio Horna Quiroz and Victor Tobon.
Like · Comment · Share · April 8

The only thing that could slightly improve this piece would have been to make sure that the copy, "It's a Jeep Thing," was more visible, perhaps by placing it onto the photo itself. With that small adjustment, Jeep would have delivered a powerful image, its logo, and its terrific tagline all in one shot. Otherwise, kudos to Jeep for such a beautiful, humanizing, and well-executed jab.

MERCEDES-BENZ: A Great Product Deserved Better

Another car company took a more traditional route than Jeep by posting a photo of their product. And what a product—that is one beautiful, luxurious car. The picture says it all, which is why it's too bad that Mercedes-Benz turned what should have been a solid jab, bordering on a right hook, into a limp poke. Here's how:

- **Too much text:** It's a shame Mercedes-Benz thought it needed to bog their stylish photo down with a load of description that few people will to read. All they had to do was include one line of text about the car's sumptuous interior, and then link out to the excellent Forbes article that told readers everything else they needed to know.
- **Poorly placed call to action:** In addition, they placed their call to action—the link to the article—at the bottom of that big paragraph of text. Why would they? Less text would have highlighted the fact that Forbes wrote such a complimentary article instead of burying it.
- **No logo:** As gorgeous as the car is, there's no way to know who made it unless you think to look at the post's profile picture. It wouldn't have sacrificed any class or sophistication to make sure the Mercedes-Benz logo was tastefully inserted somewhere on the photo itself.

Mercedes-Benz USA
What do we mean by "energizing comfort" in the upcoming 2014 S-Class? According to Forbes, the hot stone-style massage will feel like "a human hand is applying pressure." Your sumptuous leather seat, a technological showpiece in and of itself, will be heated and ventilated with 14 separately actuated air cushions in the backrest. These are but a few examples of the S-Class technology that Forbes says will set standards for the entire industry.

Read the full review and let us know what you think:
http://mbenz.us/17bxhff
Like · Comment · Share · April 5

SUBARU: Amateur Night

There is so much to dislike about this piece of content it's hard to know where to start.

- ★ **Boring text:** Like Mercedes-Benz, Subaru posted this piece to share a great review of their new car. But whereas Mercedes-Benz talked too much, Subaru has said too little. The copy length is ideal, but there was no reason to skip the opportunity to hint that the review was a positive one. What's the big secret? They've missed a chance to get the fans excited and make them want to read more.
- ★ **Terrible photo:** Unless Subaru intended to sell pavement along with its cars, there is no reason why a wet road should dominate the entire lower half of the photograph. The Subaru is so far away it's almost reduced to the same size as the little sailboats bobbing in the background.
- ★ **No logo:** There's no reason for anyone to take notice of this photograph, but even if it did somehow register, without a logo there is nothing to explain to people why this car deserves attention.

While nothing could turn this pig's ear into a silk purse, simply adding the *Consumer Reports* headline, a logo, and cropping the photo differently might turn this wasted opportunity into a serviceable jab.

VICTORIA'S SECRET: Fluent in the Platform's Language

With this powerful right hook, Victoria's Secret shows that they are fluent in native content–ese:

- ★ **Dramatic photo:** Obviously it's not just the wings this model is sporting that are going to get people—men who love what she's showing, and women who wish they had what she's showing—to screech to a halt in midscroll. But Victoria's Secret made sure that the design of the photo was as captivating as its subject. The image is big and bold enough to swallow up both a PC screen and a mobile screen; the minimalist black-and-white adds drama; the hot pink script overlaid against the model's wings is as eye-popping as her cleavage and the lingerie enhancing it. They did everything they could to make sure that no one could miss this picture if it came into their News Feed.
- ★ **Good use of copy:** The text in the photo was placed close to the center, so that even if the picture were cropped because of a small mobile screen, the text would remain visible. The voice of the status update is pitch perfect, as is the length. The copy is short and direct, but that line in parentheses delivers it with a little wink, which adds the dollop of personality and humor that is so necessary to any brand's social efforts.
- ★ **Appropriate links:** After the words "Apply here," Victoria's Secret attaches a link that takes you directly to the page where you can register for an Angel Card, making it easy and fast to make the sale. Is such a self-evident move really worth praise? You'd be amazed at how many brands set up a beautiful right hook, and then link to their general website, leaving customers fumbling around as they hunt for the appropriate tab so they can make their purchase. For an example, see the Lacoste tweet on page 96.

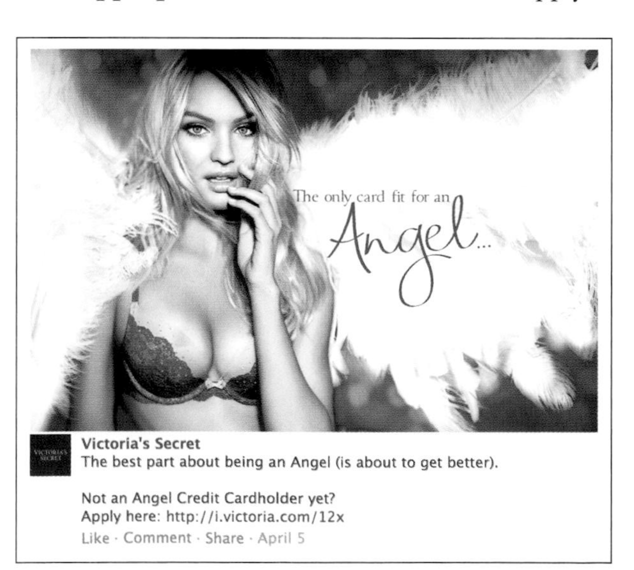

MINI COOPER: Inspiring a Spirit of Adventure

★ **Great voice:** I love the voice in this piece. In two lines, the status update promises that if you stick with Mini, you'll find adventure. You could be in Switzerland! Driving through the snow! In a convertible! The idea of driving with the top down through snow is so absurd, it's almost impossible to resist clicking on the attached link to find out how Mini could act like this drive was the trip of a lifetime. And the line "Wrap up warm" adds to our curiosity by hinting that whatever lies behind that link will put to rest any doubts we might have as to how comfortable the experience could be. Once you go to the blog post, which documents how all it takes is a pair of snow goggles and Mini's heated leather seats to make an open-air alpine drive as comfortable as a road trip down California's Highway 1, you're sold.

★ **Lacks a logo:** I'll forgive Mini for neglecting to include a logo on the photograph it used for this Facebook piece because the Mini is an iconic car, recognizable even when photographed from the back, as in this image. Still, I hope someone at the company reads this book and picks up the tip about including the logo on your micro-content, because if they start doing that their jabs will leave little to criticize.

Well played, Mini.

ZARA: Bait and Switch

With 19 million fans, Zara is a Facebook powerhouse. Why it chose to fail those fans so badly with this useless post is incomprehensible. Let's dissect why it is a complete waste of the brand's and its fans' time.

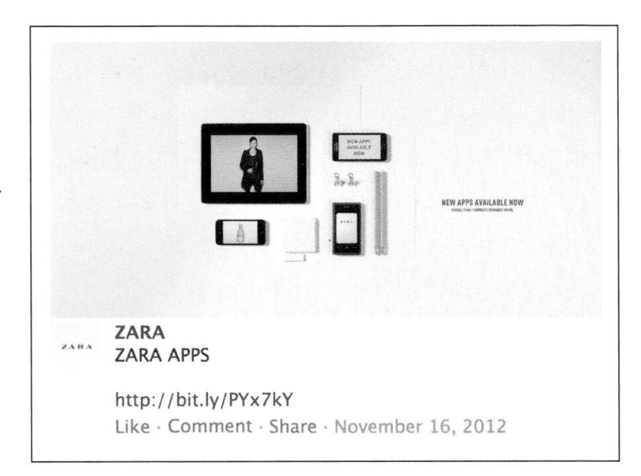

★ **Poor mobile optimization:** I had to literally squint to read the fine print underneath the headline accompanying the photos. And what the hell are those two little squiggles under the iPhone? It's even hard to make out that that yellow square is a sticky-note pad without bending your face closer to your screen. And that's when the post is viewed on a laptop! The image would have been almost impossible to see on a mobile device.

★ **Good copy:** At least they got the copy right. "Just Apps" is short and sweet and tells you everything you need to know, which is that Zara has apps. Great. Where do I get me some? Ah, a link! I'm going to click on it. Now I can . . . shop on the Zara official home page. But I wanted to download an app! Isn't that what you just announced, your apps? What the hell, Zara?

The more a brand posts links to sites that don't bring value to their customers, the more hesitant fans are going to be to click any links they see from that brand in the future. This Facebook post is a short-term fail for letting its fans down with a bait-and-switch post, and a potentially long-term fail for jeopardizing the respect and equity Zara has earned within its community.

REGAL CINEMAS: Leveraging Their Brand

No industry has a better stable of iconic images with which to leverage their brand than the film industry. Yet not long ago I was analyzing a lot of movie theaters' Facebook pages because I was considering some social media marketing opportunities, and at the time it was almost impossible to find a movie theater that used their status updates for anything other than pushing ticket purchases on Fandango. Regal Cinemas, however, bucked the trend with this successful jab that pits two movie characters against each other.

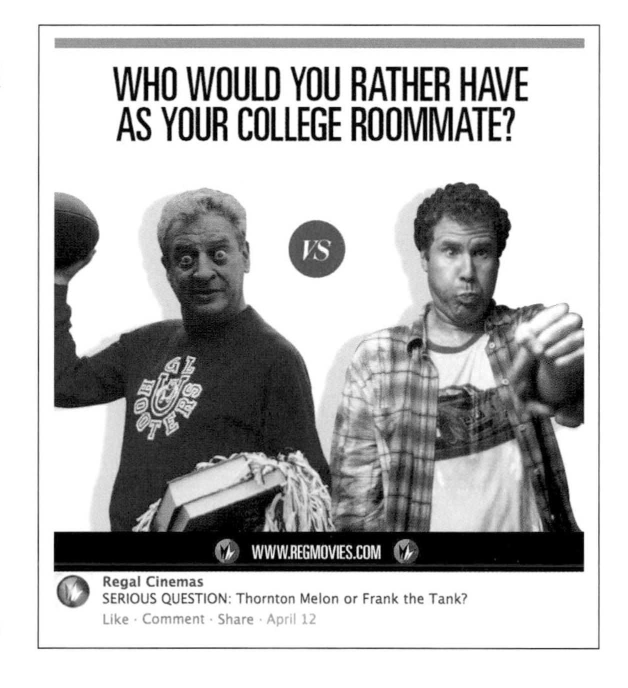

- ★ **The picture:** The theater's marketing creatives probably sifted through thousands of pictures of each of these movie characters before deciding which ones to use, and they chose well. Even though Thornton Melon and Frank the Tank went back to school in films made almost twenty years apart, they're clearly steeped in the same frat DNA.
- ★ **The copy:** For once, the status update for this content doesn't repeat the copy in the artwork. Instead, the headline of the picture sets up the question, and the status update reminds us of the characters' names, just in case someone out there isn't familiar with them. And yet, at the risk of repeating itself, the company could have seen even better engagement had they listed the names of each character under their photo, or at worst simply labeled them "A" or "B." Rule of thumb: Make it as easy as possible for your fans to engage! Why take the risk that someone won't be able to come up with the characters' names right away and therefore lose the opportunity to engage with them?

★ **Yet again, no logo:** Good for Regal Cinemas for remembering to build brand equity, but they would have been better off using a logo than a banner across the bottom of the art. Few people are going to type out the movie theater's URL, so a better use of their limited space would have been to include a sizable logo in the corner. But that's a minor criticism.

Very on point, Regal Cinemas. I'm happy with you.

PHILIPPINE AIRLINES: Totally Unappetizing

People love to talk about food, so Philippine Airlines, which flies to lots of exotic destinations, had a good idea when it decided to ask its fans to describe their most exotic meal. But after having such a good idea, why did the company waste it?

- ★ **Poor use of the platform:** It should go without saying that if you're going to talk about food, and you have the option to post a photograph, you should post the damn photograph. Philippine Airlines could have posted a gorgeous photo of a sublime Asian dish, or approached the concept with humor by photographing a plate of testicles or some other exotic dish—to Western palates, anyway—on an airline tray. It would have taken little effort to turn this content into something beautiful or fun.
- ★ **Toneless:** With airplane food the butt of so many jokes, they couldn't come up with a way to imply that Philippine Airlines knows a little bit about great food? This status update is so bland and vacuous, any company in the world could have posted it. The company simply made no attempt to make the question relevant to Philippine Airlines or its customers.
- ★ **Too many call-to-actions:** Finally, Philippine Airlines needs to remember that less is more. Doubling the number of calls to action made it more challenging to get people to answer the questions. It seems crazy, but when people are moving through the stream as fast as they do now, two questions are too much. They should have been listed as two separate posts.

SELENA GOMEZ: A Golden Touch

Your phone and your fingers are together all the time, so why shouldn't they complement each other? No wonder the hot new women's fashion trend is to match your manicure to your phone case. Here Selena Gomez laughs at herself for jumping on the fad in this savvy jab (the phone and the manicure reflect the same warm gold as Selena's promotional poster for her Stars Dance world tour), all while proving that she can carry it off to literally dazzling effect.

⋆ **The photo:** It's big and bold, very native to the Facebook platform. With Selena's glittery hand and phone swallowing up the camera, the picture would have been unmissable as fans scrolled through their news streams.

⋆ **The copy:** Celebrities are some of the worst social media abusers, and one of their biggest offenses is that they usually talk too much. Selena doesn't, and with this status update she was smart enough to keep her text short and playful.

Shared more than 6,000 times and earning more than 220,000 likes, this sponsored story with Selena Gomez shows how far fans are willing to carry a brand's content when you make them feel like everything you do is just for them.

SHAKIRA: Falling Flat

Shakira · 66,264,718 like this
April 10 at 3:40pm ·

Like

• In our video from the recent S by Shakira fragrance launch in Paris, Shak chats about being a mother, recording her next album and her role as a coach on The Voice....
• En este vídeo del viaje de Shakira a París, ciudad en la que presentó su fragancia S by Shakira, Shak nos habló sobre su nuevo rol como madre, su nuevo álbum y sobre su participación en el programa The Voice.
ShakHQ

Shakira in Paris - Shakira en París
www.youtube.com
On 27 March 2013, Shakira visited Paris to launch her new S by Shakira fragrance at the city's Sephora store. While she was there, she

Shakira rolls deep with 63 million fans, and with this post does each and every one of them, as well as herself, a disservice.

★ **Wrong type of post:** Remember how Selena's photo exploded into your line of vision? This one you have to squint to see, because it's a link post, not a photo post. When you attach YouTube links, the fill-out—the headline, link, and text—share as much space as the photo, minimizing the photo's effectiveness.
★ **Poor photo:** Not that this photo would have been particularly effective had it been any larger. The point of the post is to promote Shakira's new perfume. So why are we seeing an image of her posing with a fan and a signed soccer jersey at a podium? It's great to show how comfortable and generous Shakira is with her fans, but this is the wrong image for the purpose of this content.
★ **The copy:** First there's the copy in English. Then there's the copy in Spanish. And then there's the description in the YouTube fill-out. This isn't a novel, it's a status update, and it's supposed to be short. Brands have always been able to post according to language and location, so there was no need to double up on languages in this post.

Especially when the content is so uninspiring. It's strange that a woman with such a sizzling hot brand would post text with so little pizzazz.

★ **No engagement:** In addition, with the exception of one shout-out to her fans to thank them for liking her new Facebook page, there's no engagement between the star and her fans. That seems like a strange choice for someone who wants people to buy her perfume.

★ **The video:** It's six minutes long. No one in a Facebook mobile world has time to watch a six-minute video about your new perfume, no matter how much we like you.

The whole package, if you can stand to sit through the entire length of it, is supposed to give us a peek into the whirlwind life of a star, while revealing her humanity. There are many ways that Shakira's team could have accomplished this while bringing value to her fans.

LIL WAYNE: Welcome to Spam City, USA

There is no other way to start this review than to congratulate Lil Wayne for becoming the first person to successfully turn Facebook into Myspace.

★ **Poor page management:** Allowing people to use your fan page to build up their own businesses and Facebook pages is an insult to all the core fans who come here to be a part of your community. In addition, you risk turning those fans into antifans, as evidenced by the comments speaking up in irritation "Ok, Lil Wayne, we get it, you posted this eight times. . . ." That individual is in for a long wait if he's hoping for a direct reply, though—Weezy doesn't come here. Ever. His neglect in managing his page, cleaning out the spam, and engaging with people implies that he really doesn't care about his fans, and creates little reason for his fans to care in return, or to come back to his Facebook page.

It's tough for me to make fun of Weezy because I love his music, but honestly, when you put this little effort into your social media promotion, you're no better than the amateurs sticking promo fliers under people's windshield wipers.

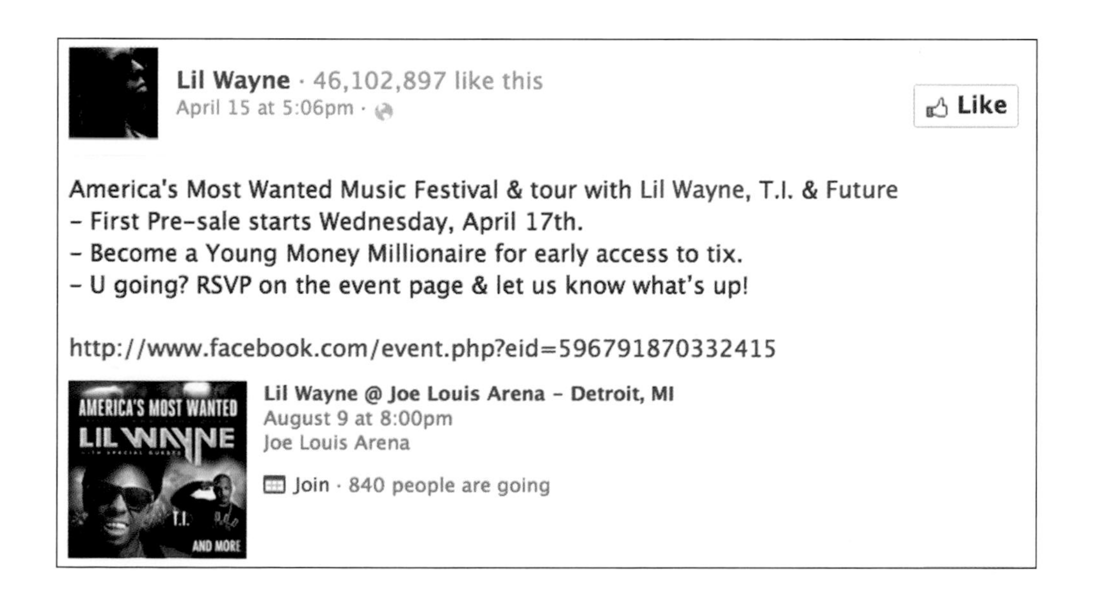

MOSCOT: Possibly the Most Confusing Facebook Post Ever

Normally, this small American business puts in a solid performance on Facebook, but this post, highlighting a positive review of the brand on an Israeli website, reveals a number of key mistakes.

- ★ **Text, text, and more problems with text:** First, there's the double copy supporting the photograph of Johnny Depp, in Hebrew and in English (though it takes some effort to find the English text). Facebook is not the place to be flooding fans with text.
- ★ **Indecipherable text:** Second, the copy that does hit us in the face is in Hebrew. It's kind of arresting, and when combined with the photo of Johnny Depp may be enough to make readers stop in their tracks. But not for long. As soon as most fans realize that they can't read anything on the page—this is an American company, and most fans will be American—they're going to move on. Few will hunt beneath the brand's tiny profile picture and click on "See More," where they will be rewarded with the English translation of the article. Last, whether in Hebrew or in English, no one should ever post copy more than a thousand words long on Facebook.

One more thing. Here and all over their Facebook page, Moscot likes its own posts. That's lamezor, Moscot. Please stop.

חדש על הקולב

המותג האמריקאי מקגרגור עושה עלייה לישראל אבל לא מביא איתו בשורה אופנתית, הנעליים של קלארקס מצטיינות בנוחות אבל מפשלות בעדכניות ומותג המשקפיים מוסקוט כבש את הסלבס אבל בהחלט לא מתאים לכולם. כל מה שחשבנו על הקולקציות החדשות שנחתו השבו בחנויות

מוסקוט

מה מוכרים לנו? את מותג משקפי הפרימיום הניו יורקי Moscot, שנכנס לראשונה לישראל. המותג, שהוקם בניו יורק בשנת 1925 על ידי מהגר יהודי והפך לסיפור הצלחה בינלאומי הנמכר ב-80 מדינות מסביב לעולם, משווק בישראל על ידי יונתן פירסטנברג, אחיה הצעיר של השחקנית חני פירסטנברג.
מוסקוט נישא על גלי התהילה של סגנון הווינטג' ומראה הגיק-שיק השולט במסגרות המשקפיים כיום, וכן על שלל הידוענים שחובבים אותו, בעיקר שחקני קולנוע כמו מישל וויליאמס, ג'ק ג'ילנהול וג'וני דפ, שלא מסיר את משקפי המותג מהחוטם היפה שלו. בעידן של תאגידי אופנה גדולים, הצליחו במוסקוט לשמור על הצביון המשפחתי של המותג, שדגמיו מעוצבים על ידי בני הדור הרביעי של המשפחה, רובם העתקים מקוריים של הבסט סלרס שעוצבו במותג מיום הקמתו ועד אמצע שנות ה-80. השמות היהודיים של הדגמים - זאב, מזל, אייזיק, קליין, מוריס, ג'ייקוב ועוד - רק מוסיפים למורשת.
האם קנינו את זה? השילוב בין מורשת ענפה של מותג, סגנון היפר-קולי ומחירים הוגנים בגזרת משקפי הפרימיום, הופכים את מוסקוט למוצר מנצח.
כמה להוציא? 1,400 שקל - מחיר ממוצע למסגרת משקפיים.

MOSCOT
Xnet
Israel
April 2013

New on the Rack
... See More
Like · Comment · Share · April ·

UNICEF: Giving Away Too Much, Too Soon

This celebrity-based post is another example of how ignoring the small nuances of a platform can make or break your content.

- ★ **Good imagery:** UNICEF did a lot right here. They had their finger on the pop culture pulse and chose the right celebrity with the ever-popular Katy Perry. The picture of a smiling Katy jumping rope with some village girls in her UNICEF T-shirt is spot-on, and should work well to bring awareness to the brand.
- ★ **Botched copy:** Where they goofed is in the copy. The first line is "Want to know what Katy Perry has been up to?" Good question. Provocative. Engaging. And UNICEF blew it by offering the answer.

The post should have ended with that first line, punctuated with a link. Leaving the question hanging would have whetted visitors' appetites for more, and kept them intrigued enough to follow UNICEF's digital bread crumbs to their website, where they could have elaborated on their humanitarian work in Madagascar and other countries. Serving up the answer right away robbed the post of all its energy and style.

It's a near miss—just one little tweak and this jab would have hit its mark.

UNICEF
Want to know what Katy Perry has been up to?

She has just been to Madagascar with us to bring attention to the situation of children in one of the poorest countries in the world, still recovering from a political crisis.

"In less than one week here, I went from crowded city slums to the most remote villages and my eyes were widely opened by the incredible need for a healthy life - nutrition, sanitation, and protection against rape and abuse - which UNICEF are stepping in to help provide."

We know that through her visit the word about the need of children in Madagascar will spread much further.

Help us doing that by sharing this post and writing a quick thank you message below or anything else you want to say!

Thank you, Katy!

© UNICEF/NYHQ2013-0169/Holt
Like · Comment · Share · April 7

LAND ROVER: Going Nowhere

I wanted to destroy this Land Rover post the first time I saw it, but as I looked under the hood, I started to wonder whether the problems plaguing this content were caused by a lack of corporate support for a creative team's honest efforts.

- ★ **No brand ID:** Don't get me wrong, the execution is weird. Imagine this coming through your stream. You see a woman peering through a telescope at you, but with no logo and no prominently overlaid text, there's no way to know what it's about unless you pause to squint at the text below.
- ★ **Wrong URL:** There, we see the post is from Land Rover, and that they've got something special planned and they'd like us to send in a passport-style photo to landroversocialmedia@gmail.com. They did a good job of keeping the text short and to the point, but then they made a surprisingly ghetto choice. Why didn't Land Rover secure a .landrover email address instead of a Gmail address? In addition, one can only hope that they aren't strict with their definition of "passport-style," because the photo they used, with half the woman's head blocked by a telescope, is not passport-style. Maybe that doesn't matter, though, because when we click through the link that takes us to a page where we can read more about the project, they don't reiterate the passport-style requirement.
- ★ **Go-nowhere link:** This error in consistency is minor, however, compared with the fact that the link takes us from the company's Facebook post . . . straight to another company Facebook post. This tells me that the creative team wasn't given the proper financial or managerial support to execute this project correctly with a proper website.

Showing off scrappy entrepreneurial spirit and making do with the resources you've got is admirable for a start-up, but not for a company like Land Rover, which sells a fairly expensive product.

STEVE NASH: A Disappointing Departure

It is entirely possible that this post was chosen for no other reason than that my dear friend Nate is a bitter Steve Nash hater for leaving his beloved Phoenix Suns, and I was only too happy to have an excuse to give Nash a negative review. That said, objectively speaking, this is one horrible piece of content.

Until now, Nash has cultivated a solid social media presence that respects the platforms and engages his fans. This piece is such a departure that it makes me wonder if he might have been surrounded by some strong social media advisers back in Phoenix, and then lost them when he moved to Los Angeles. The post was meant to promote the Steve Nash Foundation Showdown, a charity soccer match featuring NBA stars going against top *futbol* players from around the world.

- ★ **Nonnative design:** Anyone who visited Nash's fan page directly was invited to the Steve Nash Foundation "HOWDOW." If they consumed this on their phone they saw it as an "OWDOW." You've got to be smart about your status update art, and someone on Nash's media team was not.

- ⋆ **Broken link:** The URL attached to the update doesn't link out, which means Nash is counting on fans cutting and pasting the link into their URL if they want to go to the Showdown website. I assure you all of zero people did that, which is too bad because it's a beautiful website, not to mention an extremely cool and worthy cause.
- ⋆ **No spam control:** Finally, here we go again with the spam. The comment thread is littered with it. There's a plague of people who use popular fan pages to promote themselves or their businesses, and the managers of these pages need to do a better job of weeding them out.

All of these mistakes can only be a result of carelessness or laziness. Nash fans deserve better.

AMTRAK: Using Sawdust to Its Advantage

I ride Amtrak all the time, and this Facebook post made me glad that I do. I love this post—it's one of the best jabs I've seen in a long time. Best of all, it allows me to dispel some confusion about what social media can and cannot do.

★ **Great use of sawdust:** You've got to be damn smart to figure out how to take an image of something quite boring, even forgettable, like two train seats, and turn it into a fun, energizing piece of content. I call material like those train seats "sawdust"—assets that you have just lying around, maybe something you totally take for granted.

★ **Gamification:** Not only did Amtrak take advantage of their sawdust, they gamified it. Tag who you'd like to travel with—that's a fun, clever challenge that strikes an emotional punch (although it's a fairly big ask that could give you unreliable results). And what a great way to take advantage of the platform. Every person who receives a notification that they've been tagged will immediately register the Amtrak brand. It's a great way to build awareness even among people who may not already be fans.

★ **Authenticity:** There's a real person behind this post, too. You can tell because when one fan suggested Justin Bieber as his preferred seatmate, Amtrak replied with "But where would Selena Gomez go?" With one sentence, Amtrak reveals that its employees are our contemporaries, people just like us, with their fingers on the pop culture pulse, a sense of humor, and a real interest in their customers.

The only criticism worth lobbing at Amtrak is that they chose a picture of some pretty worn-out seats. The last time these seats saw some fresh upholstery was probably in 1964, when they were probably made. This brings me to the misconception a lot of marketers have about social media. It's not lipstick. No matter how brilliant, clever, or authentic you are, nothing will cover up the flaws in your content. Some people will appreciate the retro look of the seats, but a lot of people won't find them very appealing. Amtrak would have been wise to choose some less worn-out seats, or cleaned these up a little better before posting a picture of them. This poor sense of aesthetic is the only detail marring what is otherwise a perfectly executed jab.

BLACKBERRY: Missed Details Matter

My team and I struggled for several minutes to understand the story behind this post. We liked a lot about it, but then we realized that if it was that challenging to figure out what BlackBerry was trying to say, the story couldn't have registered much with an audience that probably spent less than a second thinking about it.

- ★ **Poor storytelling technique:** I understand the story that BlackBerry was trying to tell—the BlackBerry Z10 is two phones in one—one for work, one for play. And if you click on the link below the picture, you're taken to a pretty cool YouTube video that illustrates exactly what's special about the phone. In addition, you'll find another link that takes you to the product's retail site. But though the brand correctly chose to make its photo the star of the update, the image does not do enough of the storytelling for us. Why not show someone attending a kid's soccer game juxtaposed with a shot of that same person at the office? You have to look extremely closely to recognize the difference between the two screens. In addition, the text talks about work-life harmony, but the screens are reversed, in life-work order. That's sloppy. And finally, people live their whole lives looking at screens—now they have to look at screens on their screens? It's a little meta for a mobile device company.

BlackBerry was right to make a big push for this product and tell their story in social, but they should have paid more attention to the details of their execution.

MICROSOFT: Riding the Waves

It's nice to see a stodgy, unsexy company show its creative, fun side as it rides the zeitgeist.

★ **Good use of links:** In this exciting jab, Microsoft is promoting a product called Fresh Paint, an app that allows you to use a palette of colors to "paint" in templates or even on your own pictures and photographs. Fans can read all about it in the blog Microsoft posted two months before this status update, easily accessible via the link beneath the picture of Dory and Nemo. It tells us how Microsoft partnered with Disney-Pixar to create a Fresh Paint "Finding Nemo pack," a collection of original coloring pages and their appropriate palette of colors. They wisely took advantage of the announcement that there would be a sequel to *Finding Nemo* to showcase their product.

★ **Offers quality, value, and authenticity:** The post shows that the creative team at Microsoft is doing some smart thinking about where the cultural conversation is going, and how they can find ways to be a part of it. The brand receives more high marks for the quality of the image, the fact that the voice of the text isn't too corporate, and the way they brought something of value to their community. In this status update and on the blog, Microsoft really does sound excited about both the movie and their product. If only more companies would use Facebook this well.

ZEITGEIST: Missing Its Inner Hipster

This post is stunningly bad. Hipsters have told me that Zeitgeist is the ultimate hipster bar in San Francisco. Ironically, everything that's wrong with this jab could have been easily avoided if someone with a modicum of stereotypically hipster skills had created the post.

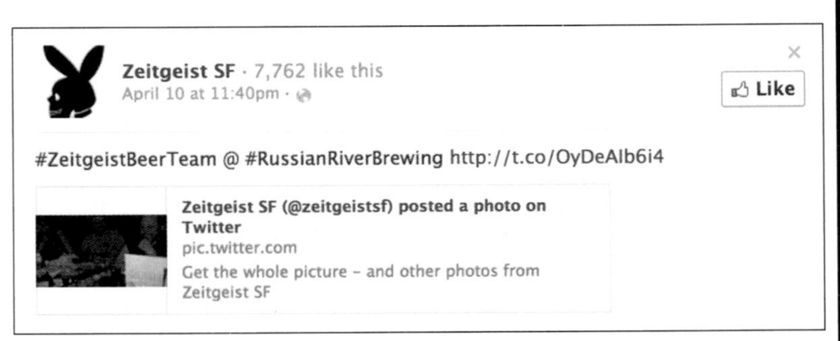

- ★ **Low Facebook value:** First, the post in and of itself has zero value but to divert fans to Twitter. There's no copy, just a mess of hashtags. Hashtags have infiltrated our culture so much that people are starting to use them as an ironic coda to status updates and even regular conversations. They have long been a huge part of Twitter and Instagram's appeal, where they overindex, and recently Facebook introduced them to the platform as well. It's possible that Zeitgeist was trying to incorporate hashtags into their voice, but they don't work here.
- ★ **Incorrect post format:** Next, it's a link post, and at the time this post was created, link posts underperformed compared with picture posts that link out (though that could change in the future). In this case, though, a picture post wouldn't have saved the status update. It might have even made it worse.
- ★ **Sorry photography:** The link takes us to a Twitter account where we see that Zeitgeist tweeted out a picture of what must be a Russian River Brewing beer tasting, showing a group of people sitting around a flight of beers. But the picture is so dark and blurry you really have to work to see what it is. That defies reason. Zeitgeist is a hip brand whose demo is all about modern technology. Photography has become a kind of social currency. This is not a great photo. It's not even good. It's the kind of picture you delete and take again. By allowing this subpar art to get posted, Zeitgeist implies that it's actually not very good at tech, and not as hip and cool as its customers. It's the kind of subliminal message that can kill a company.

TARTINE BAKERY: One Hot Mess

Tartine Bakery, a hugely popular café and pastry shop in San Francisco, has published two gorgeous illustrated cookbooks that received national attention and praise. Their Facebook post, however, indicates that like many entrepreneurs, businesses, and Fortune 500 companies, they are willing to invest energy, effort, and dollars into familiar platforms, but they have yet to put that same creative and strategic energy into the contemporary platforms where their fans actually spend more of their time. This post has so much wrong with it I have to edit my comments for the sake of space.

★ **Unclear messaging:** This post on the Tartine Bakery fan page is actually promoting an event at the bakery's sister restaurant, Bar Tartine. It's fine to cross-promote among communities, but they should have made it explicitly clear that this isn't a bakery event, since most fans are coming to this fan page looking for bakery news.

★ **Awkward text:** They write: "Bar Tartine (with Link!) hosts . . ." What an awkward, poorly written phrase. Plus, it shows that someone at Tartine actually believes that fans are too stupid to know what that little blue URL is for at the end of the post.

★ **Irrelevant hashtag:** What's up with the hashtag? The post doesn't direct us to Twitter, so what purpose does it serve here?

★ **No photo:** This is as visually unfriendly as it gets. Tartine is promoting a food-centered charitable event, and they couldn't whet our appetites with a little teaser of food porn or some other cool image to get us excited about it?

The fourth error might explain mistake number three. Not only did Tartine not include an image to accompany its charitable event, but it looks like they actually deleted one. When you attach a URL to a status update, a thumbnail image automatically appears beneath your post. But there is none here. The only way that can happen is if someone chose not to include an image. If you type the URL into your browser and head over to the fund-raising event's page, you might see why. There you'll find the most god-awful picture of a deconstructed burger ever drawn. The lettuce is vaguely dinosaur-shaped and fluorescent green; the meat, which actually looks like strips of radicchio glued together, glows red from the inside like some kind of nuclear accident about to happen, topped with fluorescent green caterpillars that are probably supposed to be pickles. It is a nightmare. No wonder Tartine Bakery didn't want that thing showing up on its fan page. Which then raises the question, why didn't they step in and provide better art to the organization that created the fund-raising Web page?

★ **Insufficient page management:** Finally, back on the fan page, the four spam comments—the only comments anyone bothered to make—are the cherries on top of this crap sundae.

TWIX: Having Fun

Twix threw a good jab here. They left their logo off the photograph, which is too bad, because as I have repeatedly pointed out, these images go through consumers' mobile streams so quickly it's easy for them to see a picture but not register who posted it. That said, Twix is such an iconic candy bar that most people will probably recognize what they're seeing right away, so in this case the omission isn't that big a deal.

★ **Clever storytelling, strong voice, good use of pop culture:** In the past, Twix has run television ads that played on the crisp sound of a Twix snapping in two, and in this post they're reinforcing that story by playing off the well-known "tree in the forest" philosophical riddle. It's a cute idea. The text shows that the writer has a strong feel for the brand's quirky, playful voice. The nice level of engagement the post received proves how appealing it is to customers when a brand skillfully inserts itself into the pop culture conversation to tell its story. They should be primed to respond whenever Twix gets ready to throw a right hook down the line.

COLGATE: Good Copy Gone Bad

★ **Catchy text:** "Did You Know?" in all caps works for me. Maybe I like the copy in this Colgate post because I grew up an ESPN *SportsCenter* fan. Regardless, this is an appropriately short, tight, positive reinforcement of the brand's interest in being important to a community that values a healthy, wholesome lifestyle. Unfortunately, the excellent text is attached to a picture that screams stock photo. Its generic personality zaps away any brand reinforcement the company could have engendered with its strong setup. Interestingly, the post did receive some strong engagement. I credit the good copy. The brand could have seen even more response had it just overlaid the Colgate logo and the text directly onto the picture. That might have even gone viral. As it stands, though, this post is a yawn.

KIT KAT: Timed Out

This is as good as a status update gets, except for one teeny, tiny mistake that makes a huge difference in the reach and influence of any post.

★ **Art, tone, logo, text—it's all good:** Posted the Friday before the 2013 Super Bowl Sunday, the art in Kit Kat's status update is fun and creative, and with pitch-perfect tone the picture and art lend an entertaining voice to the global conversation. In the right-hand corner, they included their slogan, which is an excellent alternative to a company logo. More brands should use their slogan and consistently incorporate it into their social media efforts. The product is prominent and cleverly used; the text, the tagline, and the brand slogan echo each other; the cultural reference is universal. The only misstep is in the timing of the post.

★ **Thoughtless timing:** The Super Bowl in 2013 featured the Baltimore Ravens and the San Francisco 49ers. Kit Kat launched this post at 6 A.M. Eastern Standard Time. In general, a 6 A.M. post is going to underindex because it only hits the early risers. Now, there were probably plenty of Ravens fans checking Facebook as they muddled through their early-morning routines, so it surely wasn't a complete loss. But what about the 49ers fans in San Francisco? It was 3 A.M. in their time zone when this post went live. Three o'clock in the morning has to be the single worst time you could post anything on social media. Even the people working two jobs to make ends meet are sleeping at 3 A.M. Hell, I'm sleeping at 3 A.M. (when my infant son lets me). No one on the West Coast was watching when Kit Kat posted this status update. This is a great example of how a brand's poor understanding of the psychology and behaviors of social media users can weaken their best efforts. In this case, it's a real shame, because Kit Kat's performance is so strong in this arena, other companies should be modeling their jabs after it.

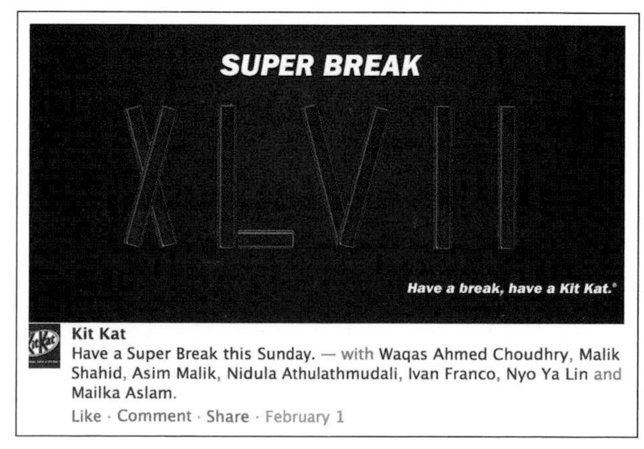

LUKE'S LOBSTER: Logoless

I love this place. Only my wife, Lizzie, knows how much—we once ate here four days in a row. On this post, Luke's Lobster did a nice job on their copy. But since the company's timeline is filled with pictures of lobster rolls pretty much 365 days a week, it would have been a nice twist to show some mother flair on their Mother's Day post. That's the missed opportunity.

The real problem, though, is that it would be easy for the speedy and casual observer to think that Cape Cod Potato Chips put out this post. Lots of brands post Facebook and Instagram shots that incorporate products from companies other than their own, and that's fine—so long as you have prominently branded your photo with your company logo in a highly noticeable corner. Which you really should do. Every time.

DONORS CHOOSE: A Solid Try

Many nonprofits litter the social media universe with such spammy content they make the likes of Lil Wayne look good (see page 59). This piece of content doesn't have the branding elements or many of the important details I've demanded from other businesses, but so few nonprofits do anything on Facebook but throw right hooks asking for money or inviting people to fund-raising galas that I wanted to give some love to Donors Choose for throwing this jab. In fact, they generally post a lot of status updates that show they're committed to jabs. I know nothing about this NGO or how it is run, but this quote seems thematically appropriate and tied to their mission. Sure, it's generic, but who knows, maybe they'll read this book and learn how to take their content up a notch. While they're at it, they can put some additional effort into their community management, which is currently almost nonexistent. If there's any place where people need to feel a strong sense of humanity, it's from the nonprofit world.

"Children must be taught how to think, not what to think."
-Margaret Mead

DonorsChoose.org
Like · Comment · Share · June 3

INSTAGRAM: A Textbook Case, and Not in a Good Way

As you'd expect, Instagram's Facebook page is filled with stunning visuals, and this one accompanying a list of Instagrammers exhibiting their work at the Venice Biennale is gorgeous. The announcement itself, though, shows that when Facebook bought Instagram, they didn't give their new employees a tutorial on how to properly storytell on their own platform. How could a subsidiary of Facebook post such copy-heavy images? There's not even a punch line or pitch. Instagram may as well have thrown up a textbook on their timeline for all the excitement that post inspires.

Instagram
Every two years, Venice, Italy, transforms into the center of the art world. Established in 1895, the Venice Biennale (La Biennale di Venezia) is a major contemporary art exhibition and the closest thing art has to the Olympics. Each of the 88 countries participating this year have selected promising artists to create elaborate installations in their designated national pavilions and palazzos. The installations are now open to the public and will be trafficked by more than 350,000 visitors before the Biennale closes on November 24.

To get a first-hand look at the Biennale, be sure to follow these Instagrammers:

* Chinese artist Ai Weiwei (http://instagram.com/aiww)
* Brazilian artist Vik Muniz (http://instagram.com/vikmuniz)
* American artist Tom Sachs (http://instagram.com/tomsachs)
* French artist JR (http://instagram.com/jr)
* Freelance journalist Erica Firpo (http://instagram.com/moscerina).

Photo by @giariv
http://instagram.com/p/Z-vf5UxqeE/
Like · Comment · Share · June 9

CONE PALACE: Yum

I need to thank Cone Palace for giving me a chance to offer an in-depth commentary at what spot-on micro-content strategy looks like. Cone Palace is an institution in Kokomo, Indiana. I can't speak about the food from experience, but if its owners pay as much attention to the quality and taste of their food as they do to their Facebook marketing strategy, it's no doubt a good reason why they've stayed in business since 1966.

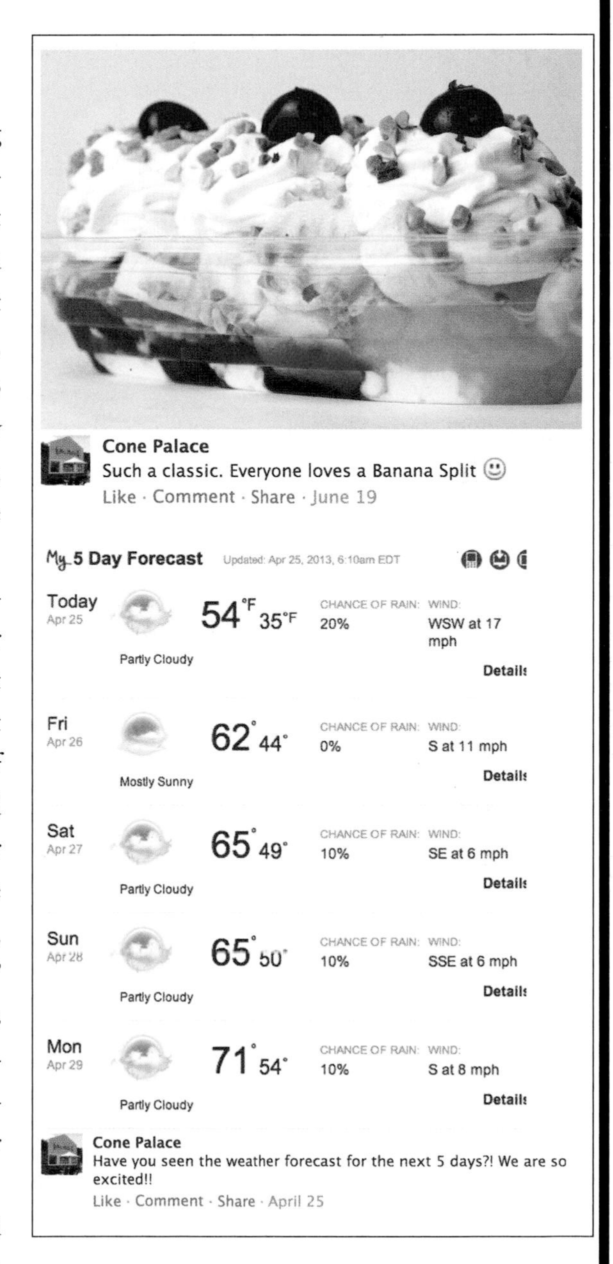

Cone Palace earned about two thousand fans as soon as they launched their Facebook page by promoting a big event and offering a 10 percent discount. But though people joined to become part of the community, they probably stayed because of the good content. Their standards are high and exacting. Before posting anything, they ask themselves, "If I saw this picture, would I share it?" If the answer is no, they don't post. That's an example many marketers should follow. Don't expect your consumers' expectations and standards to be any lower than your own.

Their posts aren't complicated, and they only put out two kinds—photographs of their food, and text posts announcing specials and new menu items, or that use local events (including people's birthdays), the weather, and holidays to provide context for their business. Hard-core analytical types might not trust Cone Palace's sometimes anecdotal, unscientific methods for measuring ROI, but when they post a picture of a hamburger and fries, and fans post comments that they're drooling and coming in for lunch, it seems safe to say that the content effectively increases sales.

And what content! Originally, staff took iPhone pictures of the food. But then they noticed that on the occasion when they had a particularly great quality photo, their engagements and interactions shot up. So they invested in a professional photographer who takes all of their food shots.

I would never have had the audacity to recommend to every business, especially a small mom-and-pop shop, that they should hire a professional photographer to take pictures of their product for social media content, because of the tremendous overhead it would represent, but secretly that's exactly what I wish every business would do. And you know, if you've got the will, there is always a way. Ever heard of bartering? That's an idea we need to take more seriously. When I think back, I could have bartered wine in exchange for professional shots of wine labels in a heartbeat if I had wanted to. If you're a small business—a shoe salesperson, a lawyer, an electrician, or maybe a real estate agent—you can provide a service or product in exchange for another service or product that you need, like professional photographs. It would be such a worthwhile investment. A beautiful photo of your product makes all the difference in the world. Flip ahead to the picture of the apple turnover on Arby's Pinterest board on page 128—would you rather eat there, or at Cone Palace?

There is one thing Cone Palace could have done better: When that generic photo of a banana split whizzed through people's newsfeed, it would have been smart for consumers to see a Cone Palace logo in the bottom of the picture or at the top left. Have I beaten that horse to death yet? INCLUDE YOUR LOGO IN YOUR PICTURE!

Kudos to a business that has figured out how to innovate and evolve for a half century, and shows no signs of stopping.

REGGIE BUSH: Being Human

Let it be said straight up that if Reggie Bush were still playing for the Dolphins, instead of the Lions, there's no way he would have made it into this book. I hate the Dolphins. But now that he's a Lion, I can give him daps. He deserves them.

Reggie Bush
If you know anything about Rowing this was my time today. Is this good?
Like · Comment · Share · April 12

Every celebrity page should be infused with this much humanity and empathy. I love what Reggie Bush has done with his Facebook timeline, offering a terrific mix of inspirational quotes, family photos, shout-outs to people he admires (both celebrity and non), and personal reflections and anecdotes. The way he tripled down on content allows him to come across as extraordinarily human. This particular photo is not perfect—the glare covers up one of the numbers. But he's using it properly to engage with his community, making it a perfect jab that will support any right hook he throws in the future.

A little Easter egg for any early readers of this book: Reggie Bush plays a Monday night game on December 16, 2013. If you read this before then, please mark the game on your calendar. After you watch, @reply me @garyvee with the hashtag #JJJRHreggiebush and give me your thoughts on the content that Reggie has been posting on his Facebook page. I will randomly pick three or four people who write to me and send them a replica of their favorite player's jersey.

Questions to Ask When Creating Facebook Micro-Content

Is the text too long?
Is it provocative, entertaining, or surprising?

Is the photo striking and high-quality?

Is the logo visible?

Have we chosen the right format for the post?

Is the call to action in the right place?

Is this interesting in any way, to anyone? For real?

Are we asking too much of the person
consuming the content?

ROUND 4: LISTEN WELL on TWITTER

- Launched: March 2006
- As of December 2012, there were more than 100 million users in the United States, 500 million worldwide.
- The Twitter concept evolved out of a brainstorming session that took place at the top of a slide on a San Francisco playground.
- The company's logo, a little blue bird, is officially named Larry, after Larry Bird, the former player for the Boston Celtics.
- JetBlue was one of the first companies to start using Twitter for marketing research and customer service.
- Users post 750 tweets per second.

I talk about Twitter with almost the same affection as I talk about my children. It's had that much of an impact on my life since I started using it to reach out to customers in 2007. As an extrovert who can get to know a room full of people in just a few hours, I felt at home in Twitter's 140-word cocktail party environment. It was the platform that came most naturally to me, because it was perfectly suited for small bursts of quick-fire conversation and idea exchanges. If the only platform I'd had at my disposal in early 2006, when I first started trying to storytell about my family business, Wine Library, had demanded long-form

writing, like a magazine column or a written blog, the business would not be what it is today. Twitter's restrictions played directly to my strengths. I owe it part of my career.

Yet discussing Twitter poses a problem for a book dedicated to improving social media content, because on this platform, and this platform alone, content often has far less value than context. How can I say that when Twitter is one of this generation's primary sources for news and information? Because with few exceptions, like the micro-content gold that is Grumpy Cat, a brand's success on Twitter is rarely predicated on the actual content it produces. Rather, it correlates with how much valuable context you add to the content—your own, and that produced by others.

Before I explain, it's necessary to acknowledge that at the time of this writing, there are changes afoot at Twitter. Until now, thanks to its origins as a mobile text-messaging service, the beauty of Twitter has been its simplicity—two or three lines of text, a link, and maybe a hashtag. But in late 2012, the company bought Vine, the six-second looping video service, and innovations such as Twitter Cards now allow people to attach photos, videos, and music directly to their tweets, thus incorporating the advantages of other more visually exciting platforms like Facebook and Pinterest. These visual enhancements will pave the way for businesses to deliver content in ways that are fresh and unique to Twitter. For example, you could tweet out a puzzle piece and announce that if one thousand people retweet it, you'll tweet another piece of the puzzle. Once all the pieces are tweeted, the puzzle would reveal where people could go to get a twenty-five-dollar gift certificate. It will be fun to explore new ways to creatively execute jabs and right hooks in such a mobile-friendly, colorful medium.

But that's all still in the works. And I'm not even sure the Facebookification of Twitter will make that much difference to those brands that haven't already gained traction there, because the additional bells and whistles won't force marketers to change how they actually use the platform. Hopefully this chapter will, though.

The main mistake most marketers make is to use Twitter primarily as an extension of their blog, a place to push a link to content they have posted elsewhere. They'll also often use it as a place to brag, especially by retweeting favorable things people say about them, a new form of humblebragging I call a "birdiebrag." There is a time and place for both of these types of right hooks, but not to the extent that most companies rely on them. Twitter primarily rewards people who listen and give, not those who ask and take. Much of the time, to read a Twitter feed is to read a mind-numbing number of right hooks. Yet if there has ever been a platform where engagement and community management have power, it's this one. There's a lot of talking and selling on Twitter, but not enough engagement, and that's a travesty, because Twitter is the cocktail party of the Internet—a place where listening well has tremendous benefits.

SPIN YOUR STORY

If Facebook's main currency is friendship, Twitter's is news and information. Go on Twitter and you'll see eighty-five people and brands at one time announcing that Brangelina is pregnant again or there's been another tornado in Oklahoma. Anyone can present news, and on their own, your tweets about your product or service are tiny drops in the deluge of information that hits people when they come to the site. The only way to differentiate yourself and pique people's interest is through your unique context. Breaking out on Twitter isn't about breaking the news or spreading information—it's about deejaying it. News has little value on its own, but the marketer who can skillfully spin, interpret, and remix it in his or her own signature style can often tell a story that is more powerful and memorable than the actual news itself.

For example, if you're a movie theater in Minneapolis, you could tweet "Just in—a great review of Bradley Cooper's newest movie from the *Star Tribune*." This is a common way to tweet—a little content, a link to a website, and you're done. But what if you put a little more than the bare minimum of effort into that jab? What if instead of offering the boring facts, you offered something fresh? How much more interesting would it be if you tweeted "The *Star Tribune* has lost its mind. This movie stinks!" and then add the link. Now that jab has some muscle behind it. Is it possible that panning something you sell will hurt sales? On Wine Library TV, I gave poor reviews to plenty of wines that were on sale in my store, and all it did was give people more reason to trust me. But if you were that worried about it, you could turn your negative review into a positive opportunity with a tweet like "The *Star Tribune* loves the new Bradley Cooper thriller. We think this movie stinks. Read. Watch. Debate." You'd then link to your blog, where you would not only have a copy of the review, but information on where and when your movie club meets every month. That's a terrific right hook. You've now positioned yourself as the opinionated, provocative movie theater that offers a unique film-watching experience, and that's a story that people will be interested in following.

Today, entertainment and escapism are prized above almost anything else. Consumers want infotainment, not information. Information is cheap and plentiful; information wrapped in a story, however, is special. Brands need to storytell around their content to make it enticing, not just put it out for passive consumption like a boring platter of cubed cheese.

EXPAND YOUR UNIVERSE

Make a statement, stake out a position, establish a voice—this is how you successfully jab your Twitter followers. But what about all those people who have never heard of you? How are you jabbing them?

Other than the easy mobile experience it offers, Twitter stands in a class apart from other social media because of the open invitation it gives us to talk to the world at large. On Facebook, Tumblr, or Instagram, you have only two options if you want to meet new fans and potential customers. First, someone might find you offline through a class, a book, an ad, or a brick-and-mortar store, and decide to follow you. Second, a customer might share a piece of your content, and his or her friend might see it and become intrigued enough to follow you. Either way, you're stuck waiting outside until that person decides to let you in. Even Facebook's search engine, Open Graph, only allows you access to stories and conversations that have been publicly shared. Everyone else is off-limits.

Twitter users, however, have an open-door policy (except for a very limited number of private profiles)—they use the platform knowing their tweets are public. In fact, that's the draw. People on Twitter are looking for attention; they welcome the spontaneous conversations that can ensue from a tweet. Strangers from around the world, many of whom will never meet in person, have been able to build robust online communities based on nothing more than a mutually shared interest in seahorses or wrestling. And people love how Twitter has allowed companies to enhance their customer service. If they want to get any brand's attention, all they have to do is mention its name and they'll get a response, because that brand is out there, using Twitter to help it communicate with its customers and build community.

Actually, that last bit is wishful thinking. Many companies are still only halfheartedly paying attention to the online conversations people have about them, thus relinquishing control over how their brand is perceived and allowing the competition to step in and shape the conversation in its favor. Fortunately, there's a book available that offers detailed explanations of why and how Twitter can be one of a business's most powerful customer service tools. It's my last book, *The Thank You Economy*. Read it, it's good.*

All (half) kidding aside, Twitter is a marketer's dream come true because it allows you to initiate a relationship with your customer. It's still the only platform where you can jump into a conversation unannounced and no one thinks you're a stalker. Here, you don't have to wait for anyone to give you permission to show how much you care. At any time, you can use the powerful Twitter search engine to find people who are talking about topics related to your business, even if only tangentially, and respond, adding your perspective and humor—and context—to the conversation.

It wouldn't take much imagination for an office furniture retailer to engage with people who mention the company name, or words like *work*, *employee*, *employer*, *office*, *desk*, *Aeron*, *printer*, *scanner*, and other office-related terms. Think of all the interesting ways it could engage with people with these words on their mind, however: *deadline*, *backache*, *fluorescent*, *happy hour*, *raise*, *promotion*, *weekend*, *swivel*, or *clutter*.

Using Twitter Search this way helps you find storytelling opportunities with people who either already know about you, or who have expressed interest in topics related to your product or service. But what about all those consumers out there who would love you if they only knew you existed? Twitter makes it possible to reach them, too. You just have to know how to ride the cultural zeitgeist.

* How's that for a right hook?

TRENDJACKING

In this chatty, 24-7 online culture, there is no better resource than Twitter trends for creating the real-time context as well as the up-to-date content so imperative to staying relevant. Twitter's trend-tracking ability is one of social media's most powerful yet underused tools. You can set your account to track worldwide, national, or even regional trends. Learning to jab with trends gives you tremendous power. You can tailor content to any situation or demographic, you can spark interest in your product or service among people outside your core group of followers, and you can scale your caring. Best of all, you can piggyback on other people's content, giving you a reprieve from having to think up fresh creative day after day. You'll still put out original content, but in this case, your content is the context you use to tell your story.

The night before I began writing this chapter, the television show *30 Rock* aired its series finale. When I went to Twitter the next day, as I expected, there it was in the list of top-ten trending topics for the United States. It seemed to me that if consumers felt like talking about *30 Rock*, marketers should be scrambling to tell their story within the context of *30 Rock*, too. Could talking about a defunct television show really help you sell more candy, crowbars, or cheese puffs? It could if you're creative enough. If you were a brand trying to ride the *30 Rock* wave, the trick would be to look for the unexpected connections, not the obvious ones. Here's one: seven. The show aired for seven years. Has your company been in business for seven years? Do you hope to do something for seven years? Do you have seven in your company name? One brand does: 7 For All Mankind, maker of premium denim clothing—sometimes nicknamed "sevens"—often worn by Hollywood celebrities. Curious to see how the brand capitalized on the Twittersphere's free gift to their marketing department, I decided to check out their recent tweets.

A look on the 7 For All Mankind (@7FAM) Twitter page the day after the end of *30 Rock* revealed some light customer engagement—which is more than some companies manage, so kudos to them—a number of retweets sharing the nice things people have said about them or their clothing line—not so great, because that's birdiebragging, and too many brands are doing it—and a stream of traditional right hooks, such as "Love a good leather tee," with a link to their product page. But nowhere was there any indication that the brand had a clue about what was going on outside the world of fashion. It was a little ironic—is there any other industry that lives for trends like fashion? One of the most successful television shows of the decade just finished a seven-year run, and 7 For All Mankind didn't even mention it. What a waste. They can talk to denim lovers every day, but on this day they had a perfect opportunity to tell their story to people who weren't even thinking about denim, and they let it pass. More distressingly, they seem to be letting all of these opportunities go by. They didn't just opt not to ride *30 Rock*; their Twitter stream revealed that they weren't riding any news or current events, except the ones they created themselves through sweepstakes, giveaways, and sales.

7 For All Mankind is a booming company that sells a great product or it wouldn't have the cult following it has garnered in the decade it's been in business. And although its Twitter profile is lacking in cultural relevance, the brand does make a serious effort to engage with its followers and stay on top of the conversation around its product. But that's Twitter 101, à la 2008. By now they should be doing much, much more. It's fortunate that the company is such a fashion powerhouse (which is also why I thought they could handle a little constructive criticism); if it were smaller and just starting out, a habit of ignoring all the opportunities to tell its story outside the parameters of denim or fashion could hurt it. Consumers don't live in a fashion bubble; why should a clothing company?

PROMOTED TWEETS

Creating context around trending hashtags only requires an investment of time, but buying a promoted tweet can be a great investment, too. On the same day that *30 Rock* trended, so did #GoRed, because the American Heart Association sponsored National Wear Red Day to raise awareness around the fight against heart disease. Above the hashtag, there was an ad for Tide laundry detergent saying, "It's crazy how Tide gets rid of tough stains, but what about the stains you want to save?" Aha. Color. With #GoRed, Tide saw an opportunity to bring attention to its color-saving capabilities. That's a clever use of a hashtag.

It was micro, it was inexpensive, and it made an impression. Think about that. Consumers are spending 10 percent of their time on mobile and there is no more mobile platform than Twitter. Yet for all the consumer attention Twitter attracts, placing an ad there still only costs lunch money compared with the price of a television ad. That was a smart use of Tide's media dollars. So many companies could have taken advantage of that opportunity. Where was Crayola? What about Target, with its big red bull's-eye? Or Red Envelope?

USING TRENDS TO THROW RIGHT HOOKS

Trending topics can be names or current events, but they can also be memes—words and phrases that have gone viral in the public sphere. These are low-hanging fruit, perfect storytelling fodder for any brand or business, especially local companies looking for a fun, creative way to differentiate themselves from their competitor.

On one of the days I was working on this chapter, the fifth-trending topic on Twitter was #sometimesyouhaveto. You can't get a better lead-in for a right hook. Literally anyone could adapt it to his or her needs:

A cheese shop could say, "#Sometimesyouhaveto eat a slice of Cabot clothbound Cheddar."

A fitness club could say, "#Sometimesyouhaveto use the sauna as incentive."

A lawyer could say, "#Sometimesyouhaveto call a lawyer to make your problem go away."

Taking advantage of hashtags is a great way for small businesses to get attention. That trending hashtag is getting clicked by tens of thousands of people. There is no reason why someone won't spot your version, like it, and go to your profile page to see what else you have to say. Once he's there, he can see the whole story you've been telling about yourself with your steady stream of jabs and occasional right hooks. He decides to follow you. Maybe he needs a lawyer. Maybe he has reason to believe that one day he will need a lawyer. Regardless, you are now that much closer to gaining a new customer when the time is right.

It could happen for a DJ in Miami named DJ Monte Carlo. While I was clicking on this trending hashtag, I spotted his tweet: "#SometimesYouHaveTo

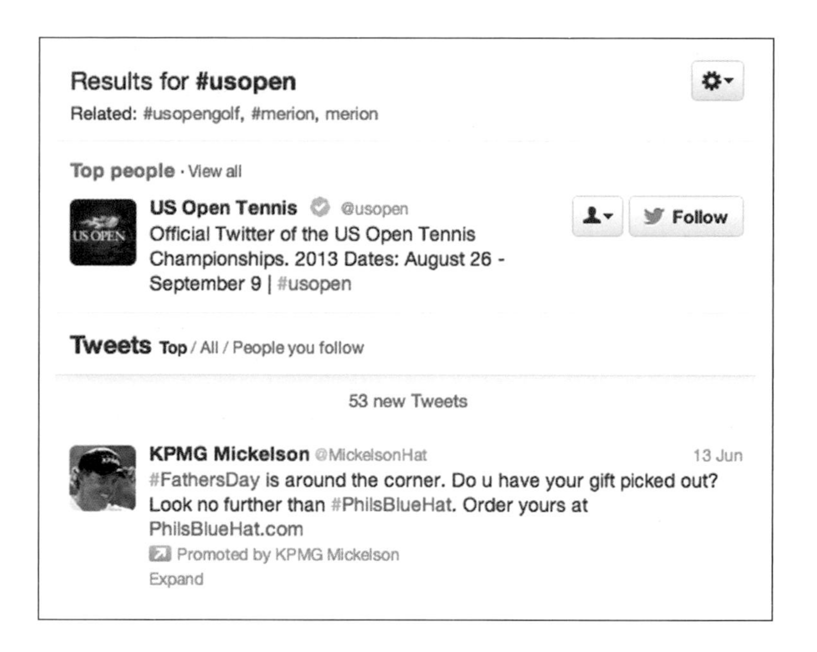

forgive those who hurt you but never forget what it taught you."

I liked that. It hit my emotional center. I decided to follow him, and he wound up in my Twitter stream, where my colleague Sam could see it. I'm not a big clubgoer, but Sam is. Maybe Sam decided to follow DJ Monte Carlo, too. And maybe, in six months, Sam will be scrolling through his Twitter feed and he'll see Monte Carlo throw a right hook announcing that he's spinning at a club in New York City that night. And maybe Sam will decide to go, too.

Get it? This is not a far-fetched scenario; it's how Twitter culture works every day. So get creative, have fun, and start experimenting with creating content on the spot, because the trending topics you see one minute will be gone the next. They have short life spans.

Something else to realize is that just because a topic is not one of the top-ten trends on Twitter doesn't mean it's not worth paying attention to. The Twitter demographic skews hip and urban, but it doesn't represent the only people talking online. You want to pay attention to what the rest of the world is interested in, too. Look for clues on Google trends. It skews young as well, just like all online data, but it reflects a broader population. During the 2013 U.S. Open golf tournament, the hashtag "#usopen" was, unsurprisingly, trending on Twitter. In response, KPMG

Mickelson, the "official Twitter account for Phil Mickelson's hat," promoted a tweet to followers of the hashtag, suggesting that golf fans honor their dads on Father's Day by donating to a charitable anti-illiteracy campaign by buying a blue Phil Mickelson hat. KPMG Mickelson didn't actually use the hashtag "#usopen" (in fact, if they're not an official event sponsor, their legal department may not have let them use the hashtag) and yet, through strategic sponsoring, they came up as the top result for anyone checking that hashtag. They were smart about the hashtag they did use, too#—fathersday.*

This example shows that KPMG Mickelson did something too many businesses don't do on Twitter: They listened. It's extremely hard to create a trending hashtag and bring people to you. It's far better to listen, find out what's trending, and bring yourself to the people. In this case, golf fans were already having the conversation. Promoting the tweet ensured that KPMG Mickelson's message became part of that conversation. It was doubly smart to include it in the Father's Day stream, as well.

United States Trends · Change
#Is1DLarryRealOrFake
#Yeezus
Father's Day
#Iran
#usopen
#ManOfSteel
Superman
MySpace
Kanye
Dad

This praise comes with two caveats:

1. Amazingly, even while KPMG Mickelson correctly joined trending conversations, they also unnecessarily included the hashtag "#PhilsBlueHat" in their tweet. How did their own invented hashtag do? A total of three people used it in the three days following KPMG's original tweet. That's embarrassing.
2. The link in the tweet doesn't actually take consumers to make a purchase. It goes to KPMG's Phil's Blue Hat website, where it takes yet another click to buy the hat. Adding extra steps after a call to action wastes the consumer's time.

Whether you jab or right hook, marketing moves like that prove that you're up

* Notice that Twitter suggests that people interested in this hashtag follow U.S. Open tennis and not U.S. Open golf. I'm not sure if this says something great about U.S. Open tennis's social media efforts or something terrible about U.S. Open golf's, or if it exploits a gaping hole in Twitter's algorithm.

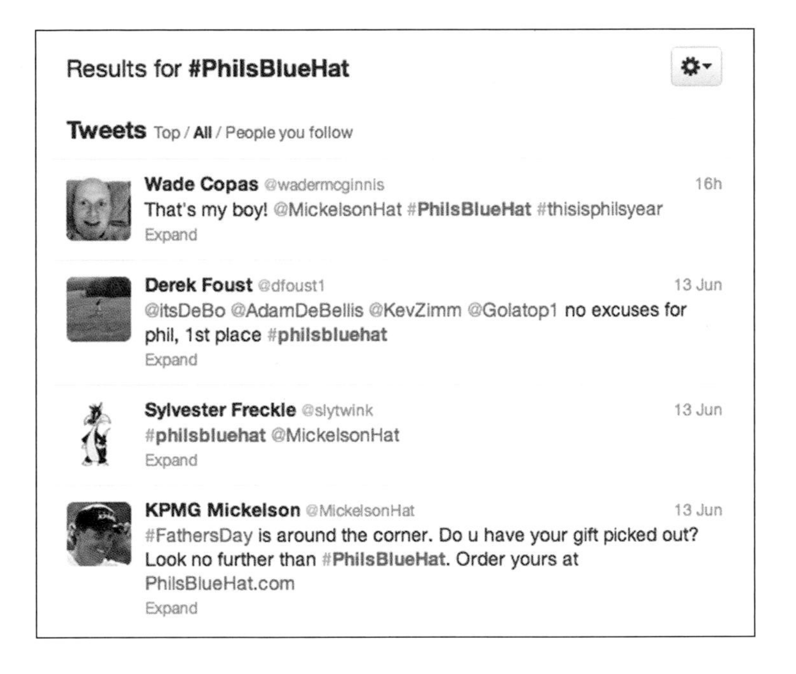

to date, that you've got a sense of humor, and most of all, that you're paying attention. You'd be amazed at how far that goes when customers are looking for someone with whom to do business.

CHOOSE HASHTAGS CAREFULLY

There's a skill to choosing hashtags. You can't just cover all your bases by tacking a bunch of hashtags onto a sentence. They won't work if they don't feel native to Twitter and natural to your brand. For example, Twitter is a hotbed of irony, but if your natural tone is generally serious and thoughtful, going ironic with your hashtags or suddenly adopting hipster vocabulary is just going to make you look like a poser. Being cool has nothing to do with age; it has to do with how solid your identity is. Do not pretend to be anyone other than who you are. That said, don't take yourself too seriously, either. Be human. If you're not comfortable talking pop culture, find someone in your organization or partner with an agency that is. Whatever you do, however, stay true to yourself. Do not pretend to be cooler than you are. Do not be the guy who hollered out, "Raise the roof!" a year too late. That's how it sounds when you use hashtags and trending topics as indiscriminate marketing tactics, instead of incorporating select ones into your conversation. Listen. Entertain, through humor or provocation.

Entrepreneurs and small businesses may see the amount of work that has to go into keeping up on Twitter and wonder if they shouldn't just give up and go home. There's no way they can compete with larger companies that have extensive budgets and staff. A person has to sleep sometime. Yes, creating real-time micro-content is an enormous job. Yes, start-ups and small businesses will have to be selective about which trends are worth their time and money. But putting your effort into that kind of thinking will do a heck of a lot more for your bottom line than sitting around like a dope waiting for customers to come to you. And it's way better than tweeting content that no one sees or cares about.

As a small business, you can gain an edge over larger companies when it comes to being nimble and authentic, two imperatives to successful Twitter marketing. Because you haven't let your personality get squashed by a PR or legal department, you have more freedom to say what you think, to look for humor in unexpected places, and to be self-deprecating. That last one works like a charm. I just admitted in an interview for *Inc.* magazine that I peed in my bed until I was twelve years old. Can you imagine anyone in a Fortune 500 company getting that personal or irreverent? Neither can I. People love it when you acknowledge your humanity and vulnerability. You may be a lightweight up against a heavyweight, but you can be the lightweight who wakes up at 3 A.M., drinks a few raw eggs, and puts in two hours at the gym before the competition's alarm clock goes off. People will notice your effort, and it will make a difference.

SCALING THE UNSCALABLE

To see what that kind of effort looks like, take a look at the conversation Levi Lentz had with Green Mountain Coffee (full disclosure: Green Mountain Coffee Roasters is a VaynerMedia client at the time of this book's publication). Green Mountain Coffee was poking its nose far beyond its comfortable coffee burrow;

otherwise it never would have seen Lentz's tweet. All Lentz tweeted was "'Say Hey' by Michael Franti is one of my favorite songs."

To his surprise, he received a reply from the verified Green Mountain Coffee Twitter account, saying, "We love that song! Isn't it motivational?"

On the surface, there is no connection between the topic of coffee and the bouncy love song Lentz was listening to. Green Mountain's jab is pure storytelling context—we're a brand that likes the same music you do. Now, what Lentz didn't know was that Michael Franti was working on a fair trade campaign with Green Mountain Coffee, so there was, in fact, a reason why Green Mountain was so interested in engaging with his tweet. However, the fact that he wasn't weirded out by the fact that a brand would contact him to talk about music proves how receptive people are to brands that reach out to consumers.

Coffee was not mentioned until Lentz brought it up, politely telling Green Mountain that he was just learning to like coffee, so he had never tried their products, but that he would definitely do so now. Green Mountain made some inquiries into his coffee tastes and followed up with a few recommendations. The conversation ended with Green Mountain asking Lentz to DM his mailing address so they could send him a Michael Franti CD, just because.

Lentz knew he was being marketed to, but he didn't care. Out of the blue, a brand had struck up an engaging conversation, given him some information he was looking for, and offered to send him a gift. Of course he wrote about it on his blog. Then he wrote about it again a few days later when he received the CD in the mail, as well as another package containing a handwritten thank-you note for writing about the company on his blog, a coffee mug, and a sample of coffee.

By watching out for opportunities to introduce itself, Green Mountain Coffee garnered extensive earned media and gained a lifelong customer by being personable, charming, generous, and above all, real, with a perfect stranger. As any good matchmaker knows, when two people are reluctant to meet, you sometimes need to find a way to firmly nudge them into the same room so they can realize how compatible they are with one another. For those companies who learn to spin compelling stories from the threads of news and information floating through the Twittersphere, this social media platform is the most indefatigable consumer-to-brand connector that ever existed.

COLOR COMMENTARY

LACOSTE: Interrupting Its Own Conversation

Lacoste is a brand with a tremendous amount of staying power. I loved Lacoste's alligator on my shirts when I was a little kid, and recently I've rediscovered the brand and started wearing it again. Reinventing yourself to your fans is no small feat, so kudos to Lacoste for pulling it off. Unfortunately, that's the only praise they're going to get from me, because this is one of the worst examples of a poorly thrown right hook in this book. It's laughably bad. I know this because I laughed my face off when I saw it.

★ **Treats the consumer like an idiot:** In the text, Lacoste asks, "If you could do one thing today, what would it be?" That's a great way to invite fans to engage. In a parallel universe, fans are posting comments like "Sleep!" "Ride a paddle boat," "Travel to Mars," "Promote whirled peas," and in all likelihood, "Shop!"—which would be an ideal moment for the brand to respond directly to that consumer and build a relationship. It would be a great opportunity for the brand to show off the personality of its fans, which in turn should reflect favorably on its own persona. But in this universe, where someone at Lacoste isn't thinking, the brand halts the conversation before it even starts by answering its own question. It's as if Lacoste didn't trust that its fans would answer the way it wanted them to. Remember, it's "Give, give, give, give, give . . . ask," not, "Give, give, give, give, give . . . demand!"

★ **Pointless link:** Like Zara on page 52, Lacoste seems to think that its website should be the hub of all its media outreach. If there's anything that brands should take away from this book, it's that there is no central hub anymore. Consumers are going to be coming through all kinds of portals, and forcing them to enter through the same door every time is going to make them tire of you. When customers click on this Twitter link, they're not taken to a special sale or even a promotion for the seasonal trends. They're just taken straight to the general website, which at the time of this writing features a blank-faced preteen.

Lacoste has more than 370,000 followers at the time of this writing. Of those followers, two saw fit to retweet this post. The link itself only received eighty-eight clicks. That's as bad as it gets. It's posts like this that are responsible for all the pointless noise on Twitter that makes it harder for the great content to get noticed. I can't even bring myself to say, "See you later, alligator," because if I see more of these kind of tweets later I may abandon the brand altogether.

DUNKIN' DONUTS: Sweet, but Out of Date

This is a charming, lightweight jab to sell iced coffee. The copy is the appropriate length, the tone is right, and the image is clever. But I have to question why the creatives at Dunkin' Donuts decided to turn their iced coffee cup into a midcentury relic.

★ **Anachronistic image:** They would have come across as a much more modern brand if they had depicted the cup with an iPhone charger coming out of it instead of a two-pronged plug that could belong to an elderly uncle's bedside table lamp. It's possible that Dunkin' Donuts purposely used an old-fashioned plug to speak to the older demographic that frequents its stores, but if that's the case, they're speaking the right language in the wrong country, because the demographic that grew up living in two-pronged-plug homes doesn't have a particularly strong showing on Twitter (three-pronged grounded outlets became a required safety feature for new homes in the early 1960s). If it is possible for "Who is Paul McCartney?" to be a trending topic on Twitter during the 2012 Grammys, then it is equally possible that half the audience that follows Dunkin' Donuts on Twitter wouldn't know what the heck that thing is sticking out of the cup.

★ **One more criticism:** The tweet is signed "JG." I understand that Dunkin' Donuts is trying to humanize their brand, but in my opinion this is the wrong way to do it. You're putting your business at risk when you let anyone except your logo or brand build equity on these public platforms. What happens when JG moves on to Starbucks or McDonald's and people start asking, "Hey, where's JG?" Your brand needs a unified front and voice. This doesn't mean you don't appreciate the efforts of the people who work for you; it means that you have to ensure that everyone is working to build up your brand equity, not their own.

ADIDAS: Slam Dunk

This Adidas Originals right hook is tremendous (yeah, the shoes are kind of whack, but . . .). I love where Adidas went with this for a few reasons.

★ **Cool picture:** They used a terrific picture of their product, clean but exploding with vibrant color. It's the kind of picture that will make a consumer scrolling through their stream stop in his tracks and take the right hook.

★ **Correct tone:** The copy is strong and builds up the story. It's written in the voice of the brand and target demo, even when they hit with the direct right hook, "Get 'em here." Often brands will write their copy with all the right slang and swag for a strong delivery, but when they go for the formal ask, that right hook, they switch to more formal corporate-speak, "You can buy them here." I love how Adidas carried the appropriate tone all the way through the right hook with "Get 'em here." Then they got right to the point, linking straight to the product page, not their home page or some other secondary page that would have required more hunting and clicking.

You want to be gentle and subtle when you're jabbing, but when it's time to ask for business, go for it. Don't be bashful. Own it.

Good job, Adidas. Very, very, very well executed.

HOLLISTER: Smart Strategy Gone Wrong

This is a really interesting case study because it represents a lot of smart strategy and a lot of awful execution all in one place.

★ **Brave creative:** Hollister deserves credit for understanding the power of Internet memes to reach a young demographic. In response to the huge popularity of planking—choosing a random location in which to lie facedown on the ground with your arms at your side—and its little brother, owling—choosing a random location in which to perch like, you guessed it, an owl—Hollister decided to try to spawn a movement toward "guarding"—holding your hands up in front of your eyes like you're holding binoculars. They went for a big right hook in asking their community to tag and engage with their meme. It's a bold move, and I love it! The problem is, though, that it's ridiculously hard for a brand to create a meme. It's not a particularly practical move, and consumers don't tend to follow it. In general, brands should be following memes, not creating them. But Hollister tried, which is admirable.

★ **Clumsy hashtag:** Where they really went wrong is in choosing their hashtag. At the time that I first reviewed this tweet, a click on #guarding showed that security guards use it, and so do sixteen-year-old basketball players. Hollister doesn't own the "guarding" concept, and so they should have chosen a more distinct hashtag to bring attention to the meme.

★ **Busy visuals:** Then there's the photo they used. It's colorful, but small and cluttered. There are too many things vying for your eye and the text is cramped. Hollister's story could have been told through a tweet in a shorter, more streamlined way with a single up-close picture of a pair of pretty boys' faces with the hashtag beneath.

SURF TACOS: Feeding New Platforms

This isn't the greatest jab of all time, but I thought it would be a good idea to show some lightweight moves that won't revolutionize the social media world but do provide some examples of easy things you can do so you don't feel pressured to create masterpiece after masterpiece.

- ★ **Good cross-pollination:** Surf Taco has a respectable following on Twitter of about 6,400 followers. They have about 500 on Instagram. By pushing an Instagram picture on Twitter, they're wisely using their bigger pool of followers to increase the size of their smaller one. This is a strategy more people need to follow, although pushing Instagram to Twitter worked better before competition between the two meant that Twitter cut off seamless Instagram integration, so that it would no longer load natively. However, when you are trying to develop a following on a new platform, whether it's Pinterest, Instagram, Snapchat, or whatever we'll see in the future, it's important to use the platform where you have the most data to drive traffic to the new one (three years ago I was telling people to use their email service to drive traffic to Facebook). Siphoning data from place to place is an excellent strategic move to build awareness of your presence on a new platform.
- ★ **Appropriate aesthetic:** Surf Taco also clearly understands the aesthetic of Instagram. This isn't a particularly artistic or exciting shot, but at least they're not using a stock photo or a glossy product picture. It's a casual, natural scene from a real place, and based on the solid engagement it received, even from a relatively small community, it resonated with followers.

They also knew enough about Twitter users to include a hashtag, and a good one, too, though it might have been smart to include one or two broad hashtags, like "#baseball," to try to earn even greater visibility.

All in all, not a bad play by a small New Jersey business.

CHUBBIES SHORTS: It's All About the Voice

Ultimately, success in social media boils down to three things: understanding the nuances of your platform, using a distinct voice, and driving your business goals. Chubbies does all three in this, one of my favorite pieces of micro-content in this book.

The most powerful thing about this piece is the voice, which carries through this content from beginning to end. It's young, wry, irreverent, and entertaining—exactly what the demo is looking for when it comes to Twitter. The tweet itself shows that the brand understands the nuances of this platform. It is brief and spare, nothing but two hashtags that link to a meme that offers humorous suggestions of things that are superior, in this case a cat named Pablo Picatso, to their competitor's product, cargo shorts. It's a ridiculous and funny comparison. Now, why did this meme work, when Hollister couldn't get much traction with #guarding? The hashtag. No one but Chubbies has any reason to create hashtags like #CargoEmbargo or #SOTO—Skies Out Thighs Out—so they have complete ownership. The hashtags are distinct enough to gain cachet to those people who decide to run with them. Chubbies didn't blow it by linking out to a product page, either.

You want to see ROI on social media? Tell a story that's good enough to get people to buy stuff. My creative team and I were impressed with this brand's commitment to upholding a strong voice and its attention to the nuances of the platform. That raised our brand awareness, which got us talking about the shorts, which made us a little obsessed, which led me to buying eleven pairs, one for each member of the team. The VaynerMedia team's thighs will be out in Chubbies style.

BULGARI US: A PR Company Gets in Its Own Way

When my parents came to this country in the late 1970s, they became obsessed with Elizabeth Taylor. In fact, I'm confident that my grandmother's first two words in English were "Elizabeth Taylor." So I have affection for the icon, which is why I hate to see her poorly treated. This was surely a great event, melding two high-end, luxury brands. Unfortunately, Bulgari didn't commit to honoring Ms. Taylor online as much as they did offline.

Live tweeting events can get obnoxious when the only value the tweets bring are to the PR company trying to get impressions. That's what's going on with this tweet. The picture is so weak, an intern hiding behind a potted plant could have taken it. We could have chosen to criticize any of the twenty-three tweets they put out throughout the day but this one deserves special attention for being particularly terrible. It's hard to even see what's going on. Try this: Turn this page back, then quickly return to this page. Can you tell what you're looking at in a split second? You have to click on the link and look on a big PC screen, and then stick your nose close to that screen, to get an idea of what the sumptuous flower arrangements on the table looked like. But no one is going to make that effort, nor should they, because the picture holds zero value, either to the consumer or to the brand.

I do give Bulgari credit for mentioning the catering company. It shows heart for an international brand to publicly acknowledge a company with a 200-person Twitter following.

NETFLIX: Simplicity Works

This is a perfectly executed jab, launched just days after Netflix announced that fifteen episodes of the long-awaited fourth season of the cult television show hit *Arrested Development* would air exclusively on their platform. Its success lies in packing a lot of power into a very simple package.

The picture is a clear reference to the show's season-three finale, when a character quits the family company. And the copy is timely and clever. "Hey Brother," a line frequently heard on the show, gave Netflix the perfect way to ride the hashtag wave of National Sibling Day. For the record, almost every day of the year has been designated an unofficial national day of something-or-other—use this knowledge well.

AMC: Calls to Nowhere

This tweet feels schizophrenic—"Retweet if you love The Rock!NO!Watch this video!NO!Buy tickets!" In 140 characters, AMC managed to make three calls to action. That's an accomplishment, but not one to be proud of. When you're asking for three calls to action, you're asking for no calls to action. The customer spotting this mishmash of links and short text coming through a mobile screen had to have been extremely confused. There's just no way to know where to focus our attention first. AMC often makes some strong social media moves, but unfortunately, much like the GI Joe movies, this one sucked.

NBA: Smart Partnering

The NBA threw a great right hook here to raise awareness of their partnership with Kia and their joint MVP awards. Every decision shows finesse, from keeping the tweet streamlined and clear, to capitalizing the word "you" to help connect with their community. They repeatedly reinforced the Kia brand, beginning with the inclusion of the Kia Twitter handle in the tweet, to framing the NBA.com landing page—which opens with an article and photo announcing LeBron James as the winner of the MVP Kia Awards—in bold red with the Kia logo. I don't know for sure that Kia paid the NBA for this fully integrated social media drive, but if they did, it was money well spent.

GOLF PIGEON: Confusing Quantity with Quality

If you're just starting out or you have a small consumer base and you want to trendjack to amplify your reach, one strategic and valuable way to do it is to use Twitter's ad platform and buy a keyword that will turn your tweet into the first or second result when a consumer searches a term on Twitter. But one thing I'm always stressing is that it's not the quantity of impressions that counts, it is the quality. You can tweet out to a million people, but if your tweet stinks or is irrelevant to them, it's entirely possible that of that million people who saw your tweet, a half million of them now hate your product or your brand. The day this tweet went out, Lionel Messi, the best soccer player in the world, must have scored his seven thousandth spectacular goal of the season, and his name was trending. Golf Pigeon must have thought that if soccer fans were talking about Messi, they might like to talk about golf, too. Wait, that doesn't even make sense. Theoretically, soccer and golf can sometimes overlap. I guess. I mean, sure, they're both sports. One explanation for this strange pairing might be that sometimes Twitter pushes promoted tweets into related hashtags to deliver more impressions. Golf Pigeon might not have chosen to promote against #messi, in which case they're off the hook. But if they did, they didn't do themselves any favors. It might have been a smart move to try to garner some crossover awareness this way back in the 1980s, when there were a limited number of channels on which to reach sports fans. But in today's targeted world, there's no reason to waste dollars marketing to a soccer community about golf. The company would have seen a lot more upside had they waited for the Masters and tripled down on trending topics that were more aligned with their brand and their community.

HOLIDAY INN: A One-Way Conversation

So many public replies, so little value. Retweeting nice things said about you to your entire consumer base has only one name. It's called bragging. Doing it nonstop is called obnoxious. From April 21 to April 23, 2013, Holiday Inn spent most of its time retweeting the nice things people said about them to all thirty thousand of their fans, when instead they should have spent five minutes forming a deeper relationship with the fans who took the time to praise them. By the way, any time a brand of this size is following more people than follow it back, it speaks to just how severely they are misusing their Twitter account. It's a sign that they're gaming the system—following people in hopes that they will follow back. It's a cheap tactic.

Poor Holiday Inn is taking the heat in this book, but retweeting fan praise is a mistake that thousands of brands make every day, probably because PR companies love to tell their clients that it's a smart move. I'm telling you, it isn't. Retweets of this nature have little to no value to anyone who follows you. It's truly poor form, not to mention incredibly boring for your followers.

FIFA: Breaking News

As I've said, businesses that want to compete in social media today need to embrace a dual identity. They will of course be the purveyors of a product or service, but they must also learn to act like a media company. This post illustrates exactly what that looks like. EA Sports FIFA is a video game for soccer lovers. But with this post, the brand shows that it understands that if it is to compete, it must become much more.

The tweet went out to announce that the teams for the UEFA Champions League semifinals had just been confirmed. Five or six years ago, soccer fans would have found this news out when it appeared at the bottom of the ESPN screen, and anyone who missed it would have read about it the next day in the newspaper. But on this day, a video game broke the news, if not to the world, then at least to anyone who followed it on Twitter. What did this jab do for the brand? The number of retweets was more than five hundred. Anyone who got their news here first turned right around and retweeted it to all of their followers. All those fans and their followers gave EA Sports FIFA the news credit. In addition, the brand reaped the rewards of nice levels of engagement, brand awareness, brand affinity, and probably tens if not hundreds of new followers—all by leading the media conversation around their genre. Those new followers represent many people who might be receptive when EA Sports FIFA throws a right hook in the form of an offer, coupon, or other call to action.

TACO BELL: Getting It

This one is impressive, a truly awesome example of skillful trendjacking. #ThoughtsInBed was trending. Taco Bell jumped in and offered their answer in their typical snarky, cheeky, edgy voice. Obviously their efforts resonated, because out of only about 430,000 followers, they received almost 13,000 retweets. Why did the tweet perform so well? Because Taco Bell did exactly what they were supposed to do—they respected the platform, and they talked in the same voice as their consumer. They understand that the Twitter demo is a youth demo, and if you look at their stream, you can see that day in, day out, they're reaching out to their followers and consistently making contact, building enormous brand affinity in the process. They deserve the highest level of praise I can offer: They get it.

SKITTLES: Hashtag Heaven

A lot of examples in this book make me want to cry, but this one made me smile. It probably made you smile, too. It's cute, it's funny, it sounds like a Skittles lover. The really smart thing they did, though, was to link their micro-content to an evergreen hashtag. It's a hashtag that never dies, its jokey, effervescent content ensuring that it remains relevant to anyone looking for a little humor. If Skittles keeps tweeting out micro-content like this, it has a long, exciting social media life ahead of it.

CHRIS GETHARD: Hard Work That Will Pay Off

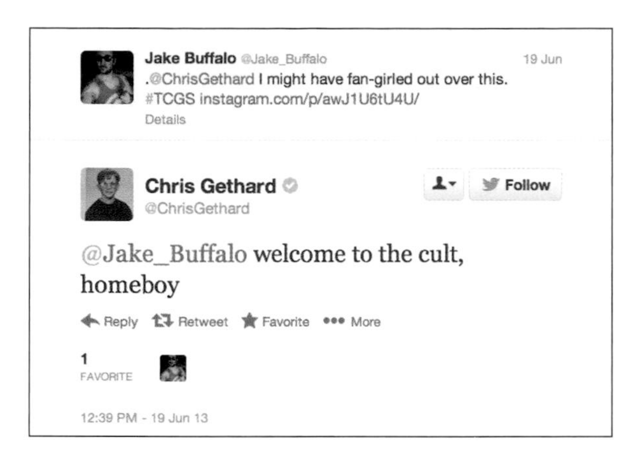

Comedians are an interesting demo on Twitter because an ungodly number of them use it to preview jokes, spread awareness, and throw right hooks like asking people to buy their DVD or come to a show. This upcoming comedian out of Brooklyn, however, has hit on the right formula. He tells jokes, of course, but he also retweets and engages, responding to and talking to fans and letting them know that he's paying attention and appreciating the time they take to let him know what they think. He's putting in a ton of effort that will result in big residual gains once he has a special or he decides to start throwing more right hooks.

TWITTER: Clueless

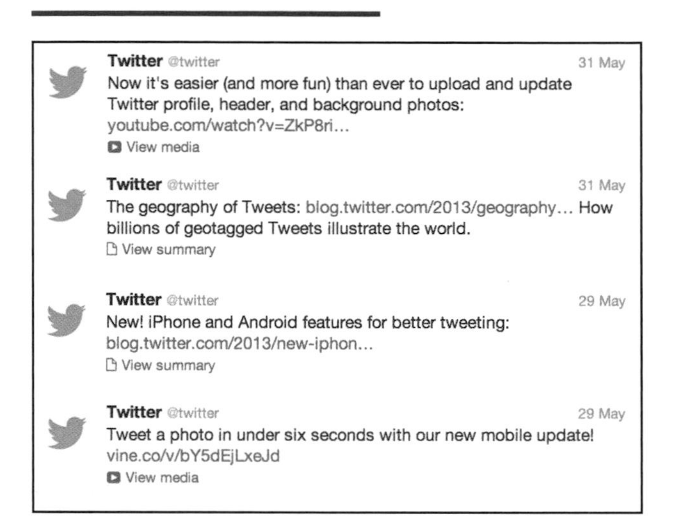

Twitter helped my career in a huge way, so it's with a heavy heart that I have to criticize them for their stunning lack of engagement. They are in a constant state of push, putting out self-serving announcement after announcement, and make zero effort to build community. On June 6, 2013, they were in full-on brag mode, announcing their new partnership with WPP. That the platform itself has no idea how to natively storytell proves that we are still living in the early days of the grand social media timeline. Twitter has the capability to listen to people talk all day long. When it first bought Vine and millions of people were tweeting raves about the new product, why couldn't it even muster up the occasional "Thank you"? How could the marketing team not realize the importance of establishing an emotional connection with its users? If they had, maybe some of the people who flocked to Instagram after it launched video sharing might have stayed loyal to Vine, instead of sending it into a downward spiral. The world is emotional. If Twitter itself is not listening and reaching out on Twitter, how can they expect anyone to feel strongly about the platform? I have a lot of friends at Twitter and I'm curious to hear their opinions when they read this critique. I'm sure they'll have plenty to say.

SPHERO: Nerding It Up

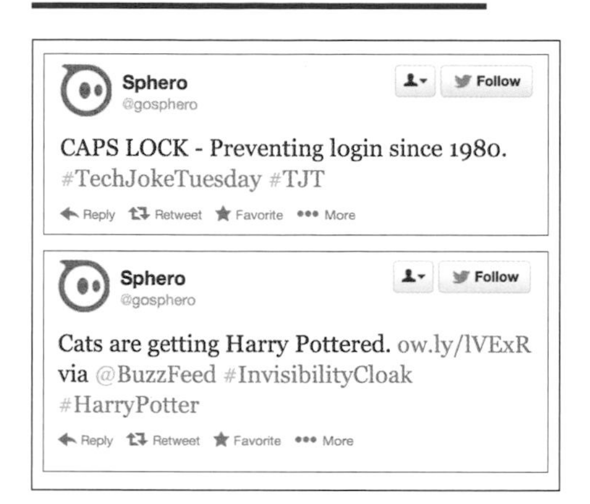

I love this in a big way. It's a perfect example of a brand that understands its audience and how to tell its story. They understand exactly who would buy a ball controlled by an iPhone. They used a video from a BuzzFeed link, which shows they speak their demo's native tongue. They get the audience, the medium, the language, the story. Even someone who didn't belong to the targeted demo would think this is cool.

Many start-ups struggle to tell good stories because instead of building community, they're focusing on fund-raising and getting an article about themselves published in TechCrunch. It's hard for a new business to strike the right balance among so many competing priorities. Sphero deserves kudos for managing to do it when so many others put it off. Truly, this is a perfect execution.

FLEURTY GIRL: Flirting with Brilliance

A lot of people reading this book are small business owners with one-store locations. Fleurty Girl has five stores, but that's still small, and the owner's commitment to her community, both online and in-store, is impressive. Born and bred in New Orleans, owner Lauren Thom throws around acronyms like *NOLA*; she knows about the peach festival in Ruston; she retweeted to a New Orleans Saints player—she's speaking the native language. She probably hasn't built up a huge base yet, but she's working hard toward it. I wish more local businesses would put her kind of energy into their media. There are ways she could add a little more spice and flair to her tweets to increase her retweet value. She could add hashtags, for example, to ignite emotions or laughs. When she tweeted, "I love peaches," an appropriate hashtag might have been #peachesfillthebelly. You need to do anything you can to get people to smile and burn a slightly deeper impression in your consumer's mind. Instead of wishing Darren Sproles a happy birthday, she could have looked up his age and matched it to the Saints player who wore that number during the 2012 season, so that the greeting becomes something a little more memorable, such as "Happy Ryan Steed!" Something like that would have been fun. I think she'll get there.

SHAKESPEARE'S PIZZA: Delicious Local Flavor

I'm happy to praise yet another small business that has made a strong commitment to putting out good micro-content, and has a talented writer creating their copy, too. Pay attention—the third tweet seems like a simple response to Earth Day, but look at the clever hashtag. That hashtag shows that this company gets the psyche of a Twitter user, that it understands that it's those little moments that make consumers go "Ha!" that compel them to retweet to friends and put your brand in their feeds. Shakespeare's Pizza could have paid for a banner ad to get an impression, but no one would have cared.

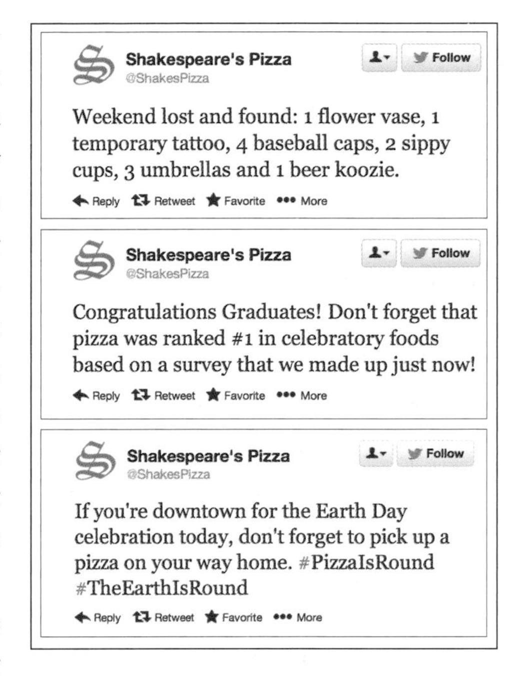

The second tweet is on point, too. Anyone between the ages of sixteen and twenty-four is going to be all in. Heck, it will appeal to anyone with the mentality of a sixteen- to twenty-four-year-old—you know who you are, hit me up on Twitter. Shakespeare's tweets prove that a combination of creative writing and a deep understanding of what brings people to Twitter will lead a brand to overindex. They also made me hungry. For the record, I like mushrooms.

Questions to Ask About Your Twitter Content:

Is it to the point?

Is the hashtag unique and memorable?

Is the image attached high quality?

Does the voice sound authentic?
Will it resonate with the Twitter audience?

ROUND 5:
GLAM IT UP on PINTEREST

- Launched: March 2010
- 48.7 million users
- Grew 379,599% in 2012
- From 2011 to 2012, Pinterest mobile app usage rose 1,698%, and users accessed the site via their mobile devices an incredible 4,225% more.
- 68% of Pinterest users are women, and half of them are mothers.
- The most repinned pin is a recipe for garlic cheese bread.

Unless you sell a product that no woman in a million years would want for herself or any person in her life—and that's a pretty limited list of products—or your legal department is dragging its feet,* you're a dope if your brand is not on Pinterest. And even if you firmly believe that you can't sell to the female demographic that outnumbers male Pinterest users by about five to one, you'd be wise to continue reading this chapter. Though the specifics of how jabs and

* Another instance in which small businesses have the advantage—no paranoid legal department to deal with!

right hooks work on Pinterest is unique to the platform, learning more about how companies successfully capitalize on the forces behind Pinterest's meteoric popularity should help fuel your creativity in devising new strategies for reaching consumers on other platforms.

Pinterest was invented to help people create online collections of things that they love and that inspire them. It immediately took off as a fantasyland for food porn addicts, fashion lovers, and people seeking home renovation and décor ideas. Then its scope quickly ballooned to reflect the myriad interests and hobbies of the approximately 48 million people currently using the site. That represents 16 percent of U.S. Internet users, only 1 percent fewer than Twitter. Yet despite its rocket ship rise to popularity, many well-established brands were slow to take it seriously. Shocker, right?

They had their reasons, of course. Part of it was probably that companies were already spinning their wheels trying to keep up with Facebook and Twitter, and just didn't want to invest in one more time-consuming social network that for all they knew was just another flash in the pan. Part of their reluctance was also probably due to early concern over the risks of copyright infringement inherent in a site that encourages people to share images they don't own. As usual, fear held big business back, leaving the terrain wide open for smaller, bolder, nimbler entrepreneurs and small businesses that were willing to experiment with various storytelling formulas on a new platform. For the record, no one has been slapped with any lawsuits. Overall, Pinterest is one giant mutual admiration society. Who is going to sue a company for pinning a picture of her product because it rocks, especially when the pin includes a link that takes consumers directly back to that product's retail page?

Now that Pinterest has revised its terms of use, has introduced business accounts, and has planned rollouts for business-friendly features, more brands are comfortable including Pinterest in their social media portfolio. Make whatever promises you must to your legal team so they can sleep at night, but afterward do not waste one more minute before creating an account so you can get your story out to the millions of people eagerly scouring the site for something new and inspiring.

PINTEREST PSYCHOLOGY 101

What's behind Pinterest's popularity with the public? It does its job well, making it easy for users to collect online research and ideas in one place on virtual bulletin boards, called pinboards, where they can "pin" images of Internet treasures they fall in love with, for safekeeping. But there's more to it than that. Pinterest also appeals to the same urge that compels teenagers to decorate their lockers with pictures of their favorite bands, office workers to liven up their workspace with bobbleheads and photos of their motorcycle trip through Argentina, home owners to place art in the middle of a window facing the street, or drivers to slap bumper stickers on their cars. We love displays and symbols and stuff that quickly and silently tells the world who we are. Better yet, we love visual reminders of who we want to be. Our homes may be cluttered, our cellulite may be out of control, and when we want to be profound we may only come up with fortune-cookie wisdom, but online, our Pinterest collections reveal that we dream of living in a serene shelter magazine spread, draping beautiful clothes over our slim silhouettes while effortlessly quoting Henry David Thoreau and the Dalai Lama. Aspiration and acquisition are two of the most powerful human drivers that lead people to buy, and Pinterest can satisfy both.

The numbers prove that the platform has become where people go to fulfill their material and emotional wish lists. A survey by Steelhouse shows that Pinterest users are 79 percent more likely to purchase something they spot on Pinterest than on Facebook. Pinterest produces four times the revenue-per-click of Twitter. Some small businesses that experimented early on with Pinterest saw as much as a 60 percent increase in revenue. Between 2011 and 2012, Pinterest's share of social-media-driven revenue for e-retailers soared from 1 percent to 17 percent.

Those statistics should send you flying to click the bright red "Join Pinterest" button to set up your account if you don't have one already. That goes for those of you who tell yourselves that your product isn't photogenic or your service doesn't translate well to imagery or it's too local. While certain platforms may be more natural fits for certain types of brands, the only limit to what your brand can accomplish on any platform is your own creativity. What's supremely fun and unique about Pinterest is that people can follow your boards, not just you the brand, which means that even if your product has some inherent limitations on Pinterest, you can still explore aspects of your brand that in other formats you might keep under wraps for fear of confusing your brand message. Pinterest gives you the freedom to set your brand's personality free.

TO BEGIN, LEARN THE ART OF THE PIN

Pinterest is eye candy, so every pin must be visually compelling. Think of your content as a collector's item. Your images need to invite clicks, and drab, boring pictures aren't going to do it. No clicks, no chance of users coming to your page, absorbing your story, and making their way into your world. Keep this in mind whether you're creating your own content or repinning content from other people's boards.

Pinterest users organize their Internet finds into categories, or boards, and businesses can arrange their content in the same way. You can use some boards to create virtual storefronts, helping users quickly and easily find what they are looking for, just as if they were in a brick-and-mortar store. So if you're a local tea shop, you could pin images under boards labeled Green Tea, Black Tea, Teas from India, Teas from China, and all the other types of tea you want to sell. You could pin images accordingly, including a price, because doing so increases the number of likes your pin receives by 36 percent, thus increasing your chances of making a sale. All pins would link back to their original source, in this case your website, so that with one click on the image your viewer can convert to a customer. It's that easy.

Yet few consumers start out their Pinterest perusals by going directly to a brand's page; they usually get there by following the images they see being repinned by others. Yet there is nothing exciting about a description like "Green Tea," and it's only a supremely dedicated green tea lover who is going to be moved to repin that corresponding image or follow that board. If anyone else does, it will probably be because you pinned a jab—something that caught a consumer's eye and compelled her to take a closer look at your page. Something like a pin with the caption "Tea You Drink After a Bad

Date," or "Tea for Handling the In-Laws," or "Tea to Celebrate Summer Break." Now you've created context, proving that you sympathize with your user's experience and that your brand has a place in her life. That's the kind of brand-to-consumer jabbing that motivates people to repin on their own boards, which exponentially increases the number of people exposed to your brand, which leads to more impressions, and more clicks to find out where the content originated, and so on and so forth down that social media rabbit hole until they land on your website, where you are perfectly positioned to make the sale with a solid right hook.

JAB TO CREATE SERENDIPITY

Many brands and businesses focus exclusively on pinning their original content, but as with Twitter, there is tremendous value in putting your own spin on the content that others bring to the platform. You may not be making direct sales, but you're offering value to consumers by becoming someone they can trust, thus increasing their incentive to come to you if they do decide they need your product or service. For example, a tea vendor may repin a picture of a beautiful teakettle under a board labeled "Tea Gear." She could then add underneath, "Pretty to look at, but be careful. Unless it's filled to the brim, you have to practically turn the kettle upside down to pour water out of it, which places your hand directly in the rising steam. We're sure the company is fixing the design flaw as we speak." You're not insulting the product, you're stating a fact based on your experience with teakettles. Or the same tea vendor could repin a picture of a tea-length cocktail dress with the description "Tea tastes better in satin." These kind of deejayed repins are the kind you want to tweet out, too. Naturally, any tweet would have the potential to bring Twitter followers to your Pinterest page, but as always, any time you invite debate and discussion or introduce elements of fun and surprise to content, you increase your likelihood of not just making a connection, but building a relationship that leads to a sale.

An effective way to attract more followers is to create boards that are only tangentially related to your brand. If all of your pins are about tea, you're only going to reach a certain demographic interested in tea. But if you created a board called "Where to Rest After a Cuppa," and pinned pictures of great hotels and other places to stay in Great Britain, India, and Asia, you'd reach a whole other category of consumers, such as vacationers, honeymooners, and business travelers. And if you did it authentically, you could even successfully create community with boards that are completely unrelated to your brand. This is where Pinterest really gives small businesses and entrepreneurs the advantage over larger organizations, because their legal and PR departments haven't smothered their personality. You can create pins about the city where you live; pins about music, books, and movies; pins about pets; pins about causes that your company supports. It's a fantastic way to tell your unabridged story, and you don't even have to say a word.

If you jab with that kind of color and creativity, people will be far more likely to pay attention to your right hooks. Among the practical lists of green, black, and pu-erh teas, and the subtle lists like Teas to Drink After a Bad Date and Teas for Sunday Mornings, you should include one aggressive sales pitch: Teas We Recommend This Month. If you've thrown enough compelling jabs, no one will find it off-putting to come face-to-face with the occasional right hook. If anything, they'll be glad you made it so easy for them to try your product.

USE JABS TO BUILD COMMUNITY

Comments are an up-and-coming aspect of Pinterest, yet they are an excellent way to instigate discovery. With so few people actively using comments on this platform to build context and awareness, it's an easy way for brands to differentiate themselves and get noticed. If you're on Twitter, you know how this works. Find opportunities to talk to people with interests that align with yours. Be genuinely interested in other people's pins and find ways to add context through conversation. By engaging with other Pinterest users, you create reasons for them to click on your name to learn more about you. Your descriptions, too, can create opportunities for other people to comment. A pin with a provocative title like "Tea You

Drink After a Bad Date" is highly likely to attract someone who will comment something along the lines of "Hope I don't need this tonight," or "Where was this when I needed it last week?" And there it is—the perfect opening to build a relationship, expand your community, and offer people something of value, if only in the form of a new, fun way to complain about the sorry state of the dating pool.

In addition, the comments give brands the chance to add their perspective to other people's pins. If the teakettle manufacturer notices that a tea vendor has questioned the design of one of its products, it should reply immediately, either explaining that the vendor is obviously misusing the kettle, or admitting the mistake and assuring the world that it is taking steps to fix the problem.

FOLLOW THE RULES

Pinterest puts a lot of energy into encouraging proper etiquette on the site, but if you think about it, the rules on Pinterest don't differ much from the rules in the real world. If you're in business, first and foremost, you have to be nice. Show your customers that you care. Exhibit your wares in an attractive and evocative way. Be generous with your knowledge. Be truthful. If you can't provide what someone is looking for, make sure to help her find someone who can. Use every customer point of contact to weave stories about who you are and what your brand stands for. Then, and only then, throw that right hook with everything you've got.

COLOR COMMENTARY

WHOLE FOODS: Feeding the Dream

More than half of the people on the site will never actually bake the three-layer cake they just repinned on their board, and an even smaller number will own a pantry like the one featured on Whole Foods' "Hot Kitchens" board. But it doesn't hurt to dream, and Whole Foods knows it. In fact, Whole Foods is a bit of a dream purveyor itself. There are probably few people who can shop exclusively at Whole Foods or who eat anything close to what might be a Whole Foods–sanctioned diet, but most of us sure would like to. With this pin and many others on its Pinterest page, Whole Foods shows that it understands that Pinterest is the conduit through which it can feed our aspirations and our yearning to live up to Whole Foods' ideals. That's why Whole Foods not only posts gorgeous images of the food we'd like to cook and eat on its Pinterest page, but it also posts pictures of the places where we would like to prepare and eat that food. Here's why this micro-content works:

* **High-quality content:** There's a reason why real estate agents and chefs don't photograph their own properties or food—no one would want it. Professional photographers know how to work the light and space to show off products at their best. The images serve as inspiration to fans, who love to imagine themselves re-creating the luxurious home interiors and dishes they see on blogs and in magazines. The fact that it would be almost impossible, since it's often the special lighting and other tricks of the photographer's trade that make the subject look so perfect, doesn't

matter. In many cases, consumers are aspiring to buy their ideal existence, not their real one, especially in real estate and food. With this repinned picture, Whole Foods successfully manages to captivate the audiences in both worlds. The image could easily be featured in an issue of Architectural Digest, and in fact it was originally taken by photographer Evan Joseph, who, according to his website, specializes in architecture and interior photography.

⋆ **Aspirational messaging:** Proving just how out-of-reach a room like this would be to most people, this particular pantry lives in a thirty-thousand-square-foot stone mansion (appropriately named the Stone Mansion) located on the former Frick estate in New Jersey. But by sharing it on the "Hot Kitchens" board, Whole Foods is essentially saying, "This is how our customers deserve to live." And that's a powerful message.

⋆ **Encourages a sense of community:** Whole Foods didn't actually create this content; it's a repin from a healthy-food and lifestyle blog called ingredients, inc. Repinning other people's material is a great way to catch potential new consumers' attention. It's also a great way to humanize your brand. It shows that you're out there reading your consumers' blogs and websites, and that you're interested in the same things they are.

⋆ **Long-term reach:** Though the "Hot Kitchens" board belongs to Whole Foods, it is actually open to at least five curators, all of whom are heavy social media influencers. In this way, Whole Foods is taking a progressive strategy by focusing on extracting the long-term benefits of collaboration and word of mouth, not the short-term boost of one-shot brand or product endorsements.

JORDAN WINERY: A Taste for Quality

Jordan Winery does a nice job of taking advantage of the functions that make Pinterest special among social media platforms:

- ★ **Aspirational, Pinterest-driven photo:** One look at the crisp, clean, magazine-worthy photograph of the wine and cheese and you start imagining yourself on a romantic date at the beach, or hosting an elegant party. The photo implies that Jordan's wine is for people with some taste, which aligns perfectly with the aspirational Pinterest demographic. It doesn't look like a winery stock photo. Rather, *Saveur* could easily have taken it during a photo shoot for a profile piece in that magazine.
- ★ **Smart labeling:** Though the photograph is meant to appeal to people with a sense of sophistication, Jordan Winery pinned it on a board called Wine 101. In other words, what they're selling is for sophisticates, but no one at Jordan Winery is a snob—the company caters to novices, too.
- ★ **Good use of links:** The image acts as a gateway to longer-form content. Clicking on the photo takes you straight to an article on the company's website elaborating on the thinking and experimentation that goes behind successful pairings of wine and cheese, as well as information about how to sign up for the tours and tastings offered on-site at the winery.

This micro-content throws a satisfying jab at both wine lovers and social media users, and for that the company gets a triple thumbs-up.

CHOBANI: Reaching the Heart of Its Users

As we've mentioned, the Pinterest audience is 80 percent female, and 50 percent of all Pinterest users have kids. With this child-centered jab, Chobani shows that it understands how to strike at the heart of the Pinterest audience.

★ **The photo:** Fun, colorful, simple. This image was chosen to make parents smile, and it probably did given the number of repins.
★ **The copy:** Fun, colorful, simple.
★ **The board:** It's smart to play toward kids, and even smarter for the brand to position itself as the source for fun, healthy snack ideas that will make mothers—and probably dads, too—feel like Superparents.

Before posting anything on this platform, ask yourself if your post could pass the Pinterest test: Could it double as an ad or act as an accompanying photo to an article featured in a top-flight magazine? If not, it doesn't belong here. For this jab, however, Chobani gets a definite yes.

ARBY'S: Sending the Wrong Message

This is as bad as it gets.

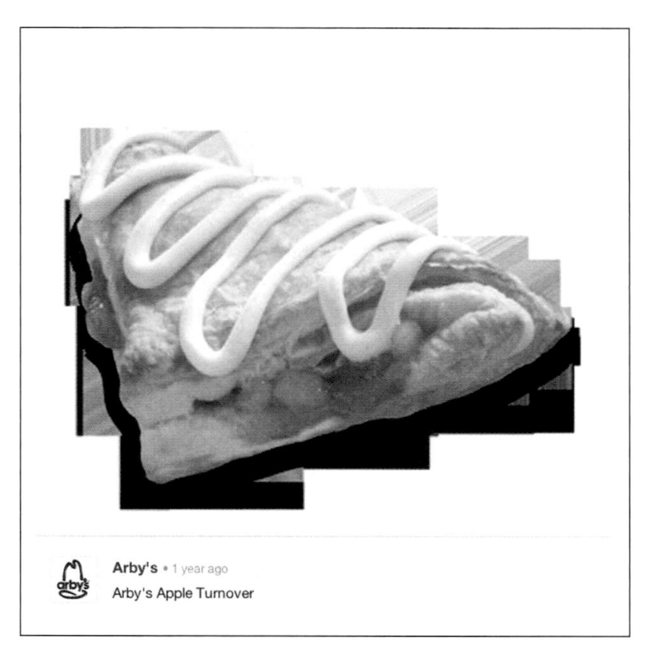

★ **The photo:** The photo itself is cropped so awkwardly, the outline of the turnover has a stair-shaped pattern to it, making it look as though the pastry is an escapee from a vintage Nintendo game where it used to threaten to smother your avatar in corn syrup and shortening.

★ **The copy:** "Arby's Apple Turnover." Wow. That's some creative text.

★ **The link:** Surprisingly, the Arby's team did know enough to link the photo to the Arby's website.

Aside from correctly linking the Pinterest post to the company website, this piece of content was a waste of the two minutes it took Arby's digital team to create. It looks like Arby's has a Pinterest account simply because someone told them they should have a Pinterest account. If they had any real interest in developing a Pinterest strategy, they would have concentrated on improving the quality of the photography and creating art that would appeal to the mostly female audience that might accidentally stumble across its board (because no one in their right mind would ever actually share this content). With an ounce of effort, they could have made this pale, pasty piece of pastry look beautiful, or at least less like something that's been sitting in a 7-Eleven display case since 1985. As it is, the only message Arby's is sending to consumers is to stay the hell away.

RACHEL ZOE: Small Mistakes Have Big Impact

Rachel Zoe provides an example of how often it's just small nuances that keep good jabs and right hooks from being great ones.

- ★ **The photo:** We see a beautiful bag, and a clear set of steps to follow to enter the Pin to Win contest. It shows creative, aggressive initiative to gamify pins and asks customers to take a social action in exchange for the opportunity to win something. The game feels authentic to the platform.
- ★ **The links:** Click on the photo of the bag, and you're taken to Neiman Marcus to make a purchase. Click on the link in the caption below the photo, and you go straight to the official rules. Someone at Rachel Zoe is thinking clearly.
- ★ **The copy:** Here's the hiccup. The copy merely repeats the three clear steps we just read in the photo. Why? With this mistake, Rachel Zoe weakened their pin's value proposition. It would have been more interesting and beneficial to customers if Zoe had added a few thoughts about the bag, and then followed up with the link to the official rules page.

In fact, what's lacking in this pin as well as on the entire board where it appears is what's lacking on a lot of celebrity Pinterest pages—the humanity. It's Rachel Zoe's name and face at the top of every pin; it would be nice to feel like Rachel Zoe had actually pinned it.

The mistakes surrounding this pin are small, but they make a tremendous difference.

Rachel Zoe • 28 weeks ago

Pin to win the metallic Eve tote from my collection by showing me how you would wear it to a holiday party on a board called "RZ Holiday Style"! Email the link to your board to social[at]rachelzoe[dot]com to be considered! Ready, set, pin! xoRZ - Official Rules: **www.thezoereport....**

BETHENNY FRANKEL: Linking to Nowhere

Bethenny Frankel, inventor of the Skinnygirl margarita mix and cocktail brand, is a heroine to every woman who loves to wear formfitting jeggings as much as she loves to drink. It's just a shame she didn't pay as much attention to the details on her Pinterest boards as she does with her product.

★ **The photo:** It's refreshing to occasionally see an unvarnished photo on Pinterest, especially on a celebrity page. You really believe Bethenny might have taken this picture herself. Normally the smeary quality is not something most people would want to associate with a food or drink product, but the picture received some solid engagement, so the DIY nature of the shot clearly didn't turn too many people off. For that, the photo gets a pass.

★ **The copy:** Skinnygirl Pomegranate Margarita. There's not much else to say, especially when a click on the picture will probably take the consumer to a recipe page or someplace fun on the Skinnygirl website. Oh . . . wait . . .

★ **The link:** When consumers link out from the picture of the pomegranate margarita, they wind up on a 404 error page, the kind that says "Page not found." That's just irresponsible. The apology offered is cute, as is the picture of the sleeping dog, but it doesn't make up for the fact that the company just wasted its customer's time and goodwill. It's a blunder that makes the brand look unprofessional.

UNICEF: Distributing, Not Storytelling

It's encouraging to know that UNICEF is progressive enough to be on Pinterest. Unfortunately, they seem to be missing the point.

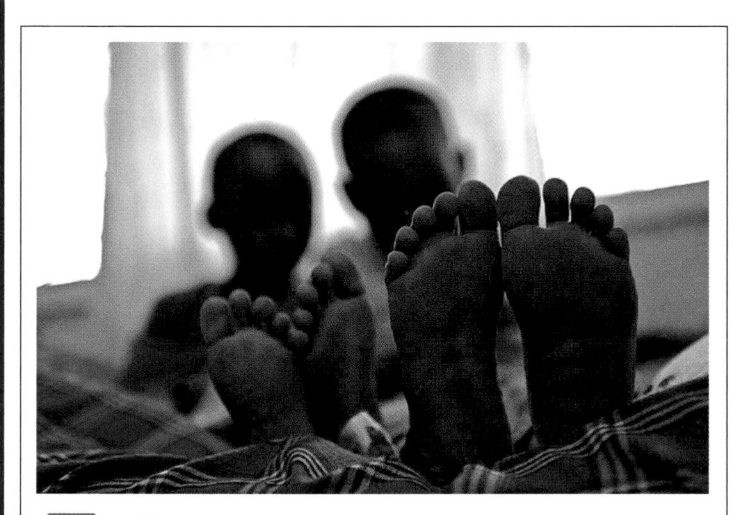

unicef

UNICEF • 12 weeks ago

CAN YOU SEE ME? --- (Left) Salome [NAME CHANGED] is HIV-positive. She is 7 years old and lives with her sister at the Turkana Outreach Orphanage in Kenya, run by Ruth Kuya, who was herself orphaned at age 12. Salome's mother contracted HIV while she was a sex worker and died of AIDS-related causes. The Orphanage began in 1994 and now shelters 40 children, most of whom have been affected by HIV/AIDS; five are HIV-positive. © UNICEF/Shehzad Noorani To see more: **www.unicef.org/...**

★ **Photo:** This piece of content illustrates a classic example of how brands mistakenly use social media platforms as distribution centers instead of storytelling venues. This photo appears on two boards. It was first pinned on one called "Can You See Me?" and then repinned on a board called "Nonprofit Media." By reposting the same photo and copy on multiple boards, UNICEF is playing for quantity of impressions instead of quality of impressions. But this strategy hamstrings the potential power of every photo on the site. It would be in the brand's best interest, especially a brand armed with as much emotionally charged content as this one, to curate boards appropriately and channel their consumers' emotions into clear calls to action. The photo would have gotten more views and more engagement if it had been posted on a board that directly appealed to people interested in helping young AIDS victims and orphans.

If UNICEF ever starts displaying its incredible photo collection with some thought to how to tell the Pinterest audience its many stories, it should start to see some impressive activity.

LAUREN CONRAD: Speaking Pinterest

Lauren Conrad's content deserves a shout-out here because it speaks fluent Pinterest. Everything about it is designed to appeal to the high-end, female audience that loves the platform. This piece could easily work as an ad or the picture accompanying an article about Lauren Conrad's workouts, and in fact, if you click on the picture, you're taken to Conrad's blog, where she suggests a workout to get your legs in shape for summer. With almost 2,500 repins, this pin shows what can happen when a celebrity brand speaks a platform's native language. This jab reflects clear respect for the platform and a commitment to her demographic. It feels right-on.

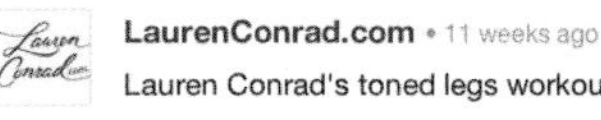

LULULEMON: Missing the Point

Once again, one mistake derailed a potentially knock-out right hook.

★ **The photo:** Infographics enjoy high levels of engagement on Pinterest, and Lululemon's gamification of the perfect yoga mat search is a creative and clever use of the medium.

★ **The link:** There is none. A click on the photo takes us to another version of the photo. Pinterest is the one place where linking out drives traffic and drives action. Why didn't Lululemon link to a retail page showing a collection of the mats described in the post so shoppers who find their perfect "mat(ch)" can actually buy one?

How disappointing to see such a fine piece of creative go to waste.

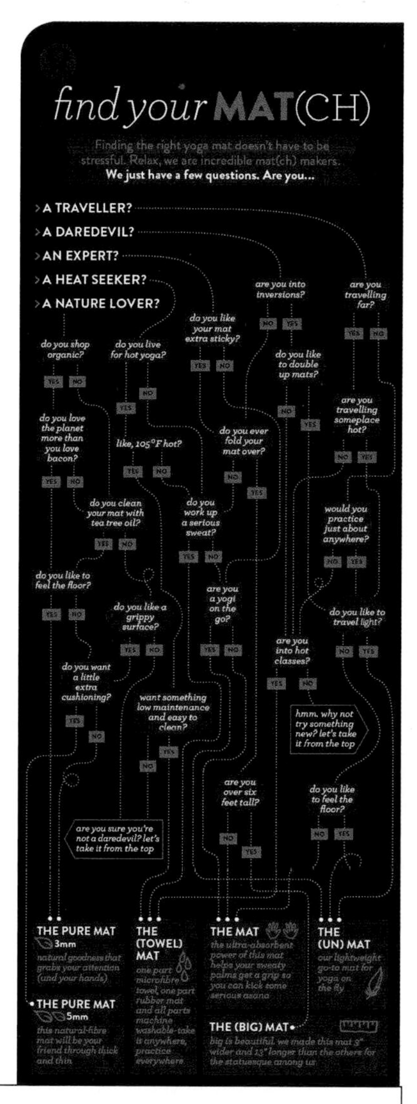

lululemon athletica • 18 weeks ago

Are you still looking for your perfect mat and aren't sure where to start? We've put together an infographic so you can find the perfect one for you - check it out!

Questions to Ask About Your Pinterest Content

Does my picture feed the consumer dream?

Did I give my boards clever, creative titles?

Have I included a price when appropriate?

Does every photo include a hyperlink?

Could this pin double as an ad or act as an accompanying photo to an article featured in a top-flight magazine?

Is this image easily categorized so people don't have to think too hard about where to repin it on their boards?

ROUND 6: CREATE ART on INSTAGRAM

- Founded: October 2010
- As of December 2012, Instagram boasted 130 million monthly active users.
- 40 million photos are uploaded per day.
- It took Flickr two years to reach the milestone of 100 million uploaded pictures; it took Instagram eight months.
- Instagram photos generate 1,000 comments per second.
- In June, 2013, Instagram launched video sharing.
- Instagram started out as a geolocation app called Burbn. When cofounders Kevin Systrom and Mike Krieger decided to revamp the app, they stripped everything out except the photo, comment, and like functions.

Instagram is another visual-centric social network. Like Pinterest, it has what I like to call "baked-in utility," meaning that it's really good at what it's supposed to do, which is help you take better mobile photographs. Yet it is a vastly more challenging platform for marketers. Unlike Pinterest, where repinning is encouraged, users can only share their own Instagram photos. And whereas on Pinterest you can embed a hyperlink into your photo that with one click will

direct users to your product or service page, Instagram is a closed loop. Anyone who clicks on your Instagram photo gets brought back to Instagram. Smart move for Instagram, not so good for marketers interested in sending traffic to a specific online location.

Given the app's limitations as a business tool, why should brands scramble to start posting photos? For the same reasons they might post ads in *Fine Cooking*, *Vogue*, *People*, or even *Traveler of Charleston* magazine. After all, if you take out the editorial content in between the ads, a print magazine is, in essence, a small-format gallery of beautiful, provocative, or tantalizing images. It's a consumption platform, and that's all Instagram is, too. It's a slightly more interactive experience than a print magazine, because users can like an image and offer comments. There's also an element of shareability and distribution in that you can connect your account to Facebook and Twitter, thus increasing awareness around your product and promoting word of mouth. Also, users can follow each other, even if they can't formally "regram." But really, when you load photos on the service, you're putting out content that no one can immediately do anything with, just like when you place ads in magazines. And you're doing it for the same reason: scale. You advertise in magazines because you know you can reach a dedicated audience, measurable by subscription rates. Instagram has incredible scale, 100 million monthly active users as of the writing of this book. With one new user joining every second, it's likely that number could increase by another 15 million by the time this book goes to press. If it's worth it to your brand to pay tens or even hundreds of thousands of dollars to place beautiful content in magazines, don't you think it's worth putting similar content on Instagram for free?

It's that scale at low cost that makes up for Instagram's lack of social value. The app's rapid growth rate proves that people are increasingly drawn to mobile, image-based content. As always, where consumers go, so should marketers. Consider Instagram as one of the great jabbing platforms, there to set the tone, tell your story, reinforce your brand, and build impressions.

Not that it's impossible to throw Instagram right hooks. Let's remember that there was no retweet option in the original incarnation of Twitter. Before Twitter developed the function, pioneers, including some of my friends and me, would share other people's tweets by cutting and pasting them into their own feeds. People are taking screenshots of photos they like on Instagram and reposting them, or using newly developed apps to the same end. There's always a work-around if you want one. You can't embed a hyperlink into your picture, but there's no reason why you can't insert a URL into your description. People aren't dumb; they'll know what to do. You could even tell people to go to your link and use the code "Instagram" to get 10 percent off your product or service (though as we've discussed, this call to action won't overindex or drive business as much as if it were linkable). Should you do this often? No, inserting too many calls to action will feel like spam. But every now and then, in the midst of jabbing, a right hook is perfectly acceptable. In fact, with so few right hooks currently in play, your right hook might be a fun surprise. But only for so long, because as we know, marketers ruin everything.

A FEW TIPS TO CREATING SUCCESSFUL INSTAGRAM CONTENT

1. Make it "Instagram." People love Instagram because of the quality of the content that has up until now been made available there. No one is going to Instagram to see advertisements and stock photos. Native Instagram content is artistic, not commercial. Use your content to express yourself authentically, not commercially.
2. Reach the Instagram generation: Learn to make Instagram work for you—it will be your gateway to the next generation of social users. The kids will be on Instagram (they're already there); their parents will still be on Facebook. I believe this as strongly as I believed back in 2011 that Facebook would buy Instagram. They did, in the spring of 2012, for a billion dollars in cash and stock. I justified the buy on *Piers Morgan* the next day, explaining that if you looked at the evolution of content from Flickr, to Myspace, to Facebook, Tumblr, and Pinterest, it

was clear that pictures were gaining in importance and were going to rule the social media world.* When Instagram started building massive momentum in 2011, there was no way that Facebook could ignore it. Despite everything Facebook had—News Feed, pages, ads—this service built on mobile and pictures posed a real threat to a company that wanted to be the best photo-sharing service around. In fact, it posed the only threat Facebook has ever faced. They had to buy it. I said that I thought the billion dollars Facebook paid was a steal, and I was ridiculed. But go figure—no one is laughing now.

3. Go crazy with your hashtags: Hashtags matter here, maybe even more than they do on Twitter. In Twitter, the hashtag can sometimes be the sprinkle—a dash of irony, a smattering of humor that you use once, maybe twice per day. On Instagram, hashtags are the whole darn cupcake. You can't overuse them. Putting out five, six, or even ten hashtags in a row per post isn't a bad way to communicate.

   And if you don't want hashtags to clutter your post copy, no problem. Put your hashtags in a comment on your photo and it accomplishes the same thing. One click on a hashtag brings a user to a whole page of other images with the same hashtag. There is no better way to earn more impressions and gain followers. Hashtags are the doorways through which people will discover your brand; without them, you're doomed to invisibility.

4. Become Explore-worthy: The most gorgeous, evocative content on Instagram gets streamed into something called the Explore page, which exposes your content to all of Instagram, not just users who have chosen to follow you. Instagram swears that the number of likes that content receives isn't the only deciding factor as to what makes it into the Explore tab, but it's surely an important one. It's a phenomenal way to build impressions. Most small businesses and even Fortune 500 brands will most likely never find themselves in this exclusive club, but any celebrities reading this book should take note of the huge opportunity.

* To watch the *Piers Morgan* clip yourself, go to bit.ly/JJJRHPiersMorgan.

COLOR COMMENTARY

BEN & JERRY'S: Sharing the Love

Ben & Jerry's micro-content is the perfect flavor for Instagram—spare and sweet. Their product delivers such a visual pop, they have no need to insert the logos that are normally an essential part of a good Instagram jab.

It's always great when a big national brand highlights one of their fans. A Swede who saw fit to post a picture of her snack prep provided this image. You can see the exchange where Ben & Jerry's reaches out to her to compliment the photo and ask permission to post it on their account at Instagram.com/ebbawallden. The only way this could have been improved would have been if Ben & Jerry's had added a virtual wink by lining up the bowl with the heart that appears when a fan likes a post.

GAP: Getting the "Social" Behind Social Media

Check out what can happen when you do your friend a favor. He works at the GAP and asks if you would use your awesome pumpkin-carving skills to carve the GAP logo. You oblige. You post a photo of your artwork on Instagram. A week later, you remember to add the appropriate tags: #pumpkin, #gap, #logo. Sure enough, you get a message from GAP asking if they can share it on their Instagram feed.

With this content, GAP shows that it really gets the "social" behind social media, and specifically, knows how to recognize material that is native to the Instagram platform. Holiday-themed content usually receives high levels of engagement, and GAP would have been crazy to skip this stellar opportunity to jab GAP fans, as well as engage with a fellow Instagram user that promoted the brand.

GANSEVOORT HOTEL: Storytelling for Love

This is a clever, artistic photo and a tremendous play. It's the kind of image that catches the heart and evokes an instant emotion in anyone going through his or her feed. Where it gets ridiculously brilliant is in its native storytelling. When you double-tap the photo, the heart shows up almost exactly in the same location as the heart on the beach. It was probably even cropped in a way to enable that action. With its smart hashtags, this is classic, fun storytelling, the kind of thing that people want to share.

LEVI'S: Blind to the Possibilities

If the goal was to permanently blind Levi's Instagram followers, this could be considered a strong right hook. Otherwise, it's really hard to tell what Levi's was trying to accomplish. It was supposed to be a creative holiday-themed piece, but holiday themes overindex because of the sense of wonder, nostalgia, or anticipation they evoke. This content doesn't evoke any emotions, nor does it tell a story, engage its fans, or do anything to enhance the Levi's brand. If this were a lightbulb company, or an electricity company, the post would make sense, but what has it got to do with a jeans company? It feels like someone got a hold of a stock photo and did what they could to make it appropriate for the holidays. This was a surprising disappointment from a business that usually does a lot to reinforce its brand.

OAKLEY: Making the Wrong Sacrifice

A visit to Oakley's Instagram profile reveals a collection of slick photographs that show off their extensive lines of sunglasses and other sportswear. But someone dropped the ball when they posted this piece of junk. And it's a shame, because the storytelling opportunity here was phenomenal.

Oakley teamed up with 2012 Masters tournament champion Bubba Watson to create the world's first hovercraft golf cart. It's an amazing piece of machinery, gliding effortlessly across the fairway, water hazards, and even sand traps, all without leaving a mark, thanks to its extraordinarily light footprint pressure. The video created to show off the invention, called "Bubba's Hover," was viewed more than three million times and received an avalanche of attention from the media. Naturally, Oakley wanted to make sure its Instagram fans didn't miss it, especially as the 2013 Masters approached.

I'm guessing—and it really is a guess—that Oakley would measure the success of this piece based on the number of views it brought to the video. That's why they lost. You can't hyperlink out of Instagram, and very few people were going to bother to highlight a link and paste it in their browser. Because Oakley was more worried about getting views of the video than crafting great content, it didn't respect the youth and creativity of the Instagram demo. They could have storytold in a way native to the platform by commissioning a cool picture of the hovercraft, maybe taken from an unusual angle, or coming up with a creative photographic teaser to entice Instagram users to make their way to the Oakley Web page featuring the video. Instead, Oakley put up a crappy still shot from the video. They got hearts, but their flat-footed execution surely meant they left a lot of engagement on the table.

THE MEATBALL SHOP:
Circumventing Instagram's Weakness with Strong Calls to Action

Right hooks are harder to land on Instagram because you can't link out, but they are possible. The key is including some really provocative storytelling in your copy to get people to respond to your call to action. The Meatball Shop understood this and made it happen. Here's how it played out.

- ★ **Start with a clever business idea: gourmet meatballs.**
- ★ **Get famous for said gourmet meatballs.**
- ★ **Take advantage of a crazy-but-true holiday: National Meatball Day.**
- ★ **Post an appropriately Instagrammy picture.**

Include a hashtag and gamify your content by urging followers to submit photos of their favorite meatball moments in exchange for the chance to be featured on the restaurant's Instagram and Twitter feeds, and receive a Meatball Shop grinder hat.

See about 1 percent of your followers engage, which is a lot for a small business with a small base.

Receive praise for a supremely well-executed Instagram right hook in a book,* which leads many more people to become aware of the shop. And to crave meatballs.

* Future editions will hopefully replace the word *book* with "number one *New York Times* bestseller."

BONOBOS: Smart Cross-Pollination

Bonobos started out as an Internet-only fashion brand, so with their roots firmly planted in digital soil it's no surprise that they show tremendous savvy when it comes to exploring the possibilities inherent in new platforms. Cross-pollinating between platforms is a great way to build brand awareness across the board, and here Bonobos shows tremendous savvy as it throws a right hook by inviting followers to preview its fall-winter line on Vine. See the proper use of and engagement with hashtag culture. See the subtle branding of including the Vine logo in the bottom right corner. See the sparse and arty look of the photo.

By paying attention to all of the details, Bonobos not only threw a successful right hook, but also perpetuated its image as a hip, creative, innovative company.

SEAWORLD: Sloppy, Sloppy, Sloppy

Sometimes when you're good, it's more noticeable when you step out of line. SeaWorld usually offers some strong, engaging content on Instagram, but not this time. You'd think a theme park would have an interest in making sure that their event seems unmissable, but this post makes it look as through attendees are in for a night with about as much entertainment and excitement as a college band reunion concert. The picture is hazy, the dates on the poster are cut off—what was SeaWorld thinking? It's bad to throw a sloppy jab, but it's even worse to throw a half-assed right hook, which is what this is. Truly, one of the worst I've seen.

GUTHRIE GREEN PARK: Acting Human

Think about the park near your home. Would you ever believe it could become a dominant presence on a social media site? Unlikely, right? Yet here's a park that is building brand equity by nimbly jabbing on its Instagram account. By regramming pictures taken by Tulsa, Oklahoma, residents and visitors to the park, Guthrie Green is acting like a real person, which makes it part of the community, and thus gives it clout. It's a brand born in social and because of that genesis it has the ability to act social. I love showing off an organization that really gets it, but more than that, I love getting a look at the future. This park will soon not be an anomaly. Every start-up, new business, and new celebrity in the future will be a native creature of the social web.

COMEDY CENTRAL: Bringing Community Together

It's a shelfie. Get it? That's freaking funny.

I've bashed others for low-quality pictures, and this one isn't spectacular, but the content as a whole is so good I'm willing to forgive. Though the quality of the photo is poor, it is highly authentic—nothing about it feels scripted. The viewer feels privy to a random, spontaneous bit of cosmic hilarity. What elevates the picture, however, is the single hashtag "#shelfie," a hashtag that plays off the mother of all hashtags that dominates Instagram, "#selfie." The pun is funny, clever, on voice, and reinforces the brand. It's the kind of content that gets shared, and shared a lot. Comedy Central really gets the power of Instagram. No matter what else may be going on in the world, Comedy Central successfully uses the platform to create a moment and bring its community together for a shared laugh. That's priceless. That's the magic made possible when a brand truly understands a social media platform.

Questions to Ask About Your Instagram Content

Is my image artsy and indie enough for the Instagram crowd?

Have I included enough descriptive hashtags?

Are my stories appealing to the young generation?

ROUND 7: GET ANIMATED on TUMBLR

- Launched: February 2007
- As of June 2013, 132 million monthly unique users
- 60 million new posts every day
- The Tumblr blog was originally on WordPress; it didn't move to Tumblr until May 2008.
- For every new feature Tumblr introduces, an old one is removed.
- Ranks number one in average number of minutes per visit (Facebook ranks third)
- Bought for $1.1 billion by Yahoo on May 19, 2013

Tumblr isn't for everyone. It skews young, appealing largely to eighteen- to thirty-four-year-olds with a slight tilt toward women. In addition, it skews extremely artsy, providing an exhibition space for photographers, musicians, and graphic designers. If Twitter is hip-hop, Tumblr is indie rock. And yet, though Tumblr doesn't have the scale of Pinterest or Instagram, you should be there.

I have a soft spot for this platform, and even invested in it in 2009. I became a huge fan during the early days of my career, both because it was so easy to use and because its minimalist format invited less text-heavy, more visually oriented posts. In fact, Tumblr's young founder, twenty-six-year-old David Karp, created Tumblr because though he wanted to blog, he found the "big empty text box" of traditional blogging platforms too daunting. His problem was the same as mine: He had tons of ideas to share, but he hated to write. Tumblr's *obstsalat* (German for "fruit salad") format became a perfect platform for the random bits and pieces of content that started getting tossed around as users scrolled through the site.

Most people continue to think of Tumblr as a mere blogging platform, but in the few short years since its launch in 2007, it has become much more. In January 2012, it debuted a newly streamlined dashboard that suggested an attempt to Twitterfy itself and embrace its evolution as a full-fledged social media site. And in a *Forbes* interview that same month, Karp referred to it as a "media network." So what is it? It's all of those things, but to get the most out of it, brands should approach it as a brandable, unique, micro-content exhibition space and sparring ring.

WHY IT'S BRANDABLE

Tumblr can't be beat as a branding platform. When selecting a background for your home page, you can choose from a series of Tumblr-designed "themes." If you wish, you can tweak those to your liking. But you can also create a completely custom look, one that perfectly reflects your brand and continues the story you're telling through your content. Color, format, font, logo placement, art—you can be as creative as you want. Unlike on Facebook, where you are locked into a definite Facebook "look," or even Twitter, where despite some profile page customization options, users are limited to seeing an endless slot-machine blur of plain text, Tumblr gives you complete artistic control. It represents the perfect opportunity for brands to experiment with new creative storytelling forms.

WHY IT'S UNIQUE

Unlike Facebook and Twitter, which guide social connections through who you know—the social graph—Tumblr was the original interest graph platform, meaning connections are made based on what people are interested in. Produce the right eye candy for your audience, and they will find you. And on Tumblr, there is a particularly tasty bit of candy at your disposal that you can't post on any other social network: the animated GIF.

The acronym stands for Graphics Interchange Format, which does little to explain what the heck they are. But you've seen them. They're so popular that the *Oxford English Dictionary* chose *GIF* as its 2012 U.S. word of the year. If you're old enough to remember *Ally McBeal*, you'll remember that dancing baby that showed up everywhere for a while. That was one of the early animated GIF memes. Today, you might see someone post a looping three-second moving image of Oprah strutting through her audience, or an otherwise still shot of a landscape with trees that blow in the wind. That's an animated GIF. People have also adopted them as live-action emoticons, using animated GIFs of celebrities with their jaws dropping open, for example, to express surprise and shock.

Animated GIFs are becoming a whole new cultural movement and vehicle for self-expression, and the best place to find them is on Tumblr. People are creating amazing art with the form, transforming ordinary images into magical mini-worlds. A picture of a fish is beautiful; a picture of a fish whose mouth is open and closing is surprising, funny, dramatic, and kinetic. You can use an animated GIF for your Twitter profile picture, but in general, aside from Google+, there is no social media site that allows you to take advantage of this gorgeous, powerful storytelling format the way Tumblr does.

Does it matter that much, especially when the scale of viewers is so much less here than on other image-heavy sites like Pinterest and Instagram? An unscientific comparison of still images to animated GIFs on Tumblr often reveals that people are driven to engage with moving pictures far more than they are with static ones. Many times, a gorgeous photo will get three times fewer hearts, or likes, than the relatively dull image right next to it, simply because the dull image is also an animated GIF. Animated GIFs are still so new that they offer an element of surprise and wonder. What's a marketer's job if not to treat your customers to surprise and wonder?

WHY IT'S A TERRIFIC SPARRING RING

Tumblr has always been more of a publishing platform than a consumption platform, but people do consume there, just at an incredibly rapid rate. That's why it's perfect for mobile: because users can just scroll and scroll and scroll and feed themselves with an endless stream of beautiful, even haunting images.

The jabbing possibilities should be obvious. Tell your story and create brand impressions through amazing art that highlights what makes your brand special. Tumblr is arty, and so is its audience. This isn't crafty, scrapbooking Middle America; this is urban loft, bike-riding, ironic eyewear America. Study the platform, figure out what people are looking for, and give it to them in the platform's native tongue, preferably in GIF form. That's your surest way of getting people to slow their speed-scroll down to a crawl, and maybe even stop to vote their approval by liking it with the little heart button or commenting with a note. Don't be afraid to deejay other people's content by adding your own copy and posting it to your blog, either. The easy shareability of content makes community building a cinch here. Make sure to add plenty of detailed tags to make it easy for people looking for content like yours to find.

While Tumblr is overwhelmingly a platform ripe for jabs, right hooks are possible. Just keep them very, very quiet. Every now and then, add a link to the bottom of your content that directs users to your Web page or retail site. If your content is as good as it should be, people will be thrilled to see that they can purchase your cool product or service. In addition, as with all platforms, keep an eye out for future opportunities to convert the sale. Even if you don't feel like Tumblr is an optimal site for you, it's better to get there early and get comfortable so that by the time your competitors recognize that they've been missing out on an opportunity, you've cornered the market.

I believe all of these tips will remain relevant even though as I was putting the finishing touches on this chapter, Yahoo bought the company for $1.1 billion. My opinion may be a little skewed since I am fortunate enough to be an investor in Tumblr, but I don't think this purchase will result in a lot of changes to the platform. Yahoo will probably take a hands-off approach and simply let David Karp's genius run unfettered. It's likely that we will see a little more aggressive advertising on the platform, but if Yahoo has any sense, it will navigate this acquisition in the same way as Facebook did when it bought Instagram—it will leave it alone.

COLOR COMMENTARY

LIFE: Successfully Bridging Generations

We've talked about how one of the biggest advantages to Tumblr is that it provides a native platform for the animated GIF, and that it's a habitat for young, hip artists and progressive companies. Yet one of the best pieces of Tumblr content represented in this book is neither an animated GIF, nor the brainchild of a particularly progressive brand. It's a sixty-year-old black-and-white photograph first published in a magazine whose name lives on only on the Web (with the exception of the occasional special magazine you'll find by the checkout register at the grocery store).

And it's freaking awesome. Here are all the reasons why.

★ **It scores high on the cool spectrum:** Tumblr demands coolness. Is there anyone cooler than Marlon Brando? Even people who have no interest in the brand's history as a pioneer in photojournalism will be captivated by this image, and curious to know more about the company that posted it.
★ **It rides the pop culture zeitgeist:** By posting this picture on Brando's birthday, when the actor was already bound to be part of the global conversation, *Life* gave it a much better chance of being noticed by consumers and other publications than if they had posted it on any other random day.
★ **The content is a rarity:** In releasing this previously unpublished photo from its archives, *Life* built its street cred as a purveyor of exclusive and elusive content, which is exactly what the Tumblr audience wants. The word-of-mouth potential is huge because consumers will share the content just so they can be the first among their friends to say they spotted it.

Life's execution of this content was spot-on, and continuing in this vein should help this old-school brand build some recognition and give it access to the younger generation.

PAUL SCHEER: Storytelling in Place of Self-Promotion

You've seen Paul Scheer before, you just didn't know it. He's the B-list—actually, that may be generous—he's the solid C-list comedian with the Grand Canyon–sized gap in his teeth who has appeared in everything from the police procedural parody *NTSF:SD.SUV* on Cartoon Network's *Adult Swim* to *30 Rock* to *Yo Gabba Gabba*, and currently costars in the fantasy football comedy *The League* on FX. Obsessed with AMC's drama *Breaking Bad*, Sheer created a Tumblr blog to spread the word about the show to his fans and make sure they start watching. In doing so, of course, he also gave the general public a reason to start watching him. And they should, because he's brilliant.

★ **Smart use of native content:** Scheer takes advantage of the only platform that gives him access to the medium that most overindexes with social media users, the animated GIF, going so far as to herald it as the next ascendant art form. "If Leonardo da Vinci painted the Sistine Chapel today, he would do it with GIFs." (I know, Michelangelo painted the Sistine Chapel. But he would use GIFs, too.)

★ **Takes advantage of pop culture:** *Breaking Bad* is hugely popular with its fans, so rather than try to compete for their attention, Scheer simply used the Tumblr platform to make himself part of a conversation that was already happening.

★ **Promotes the brand rather than sells the brand:** Rather than doing obvious self-promotion, Scheer uses the blog to storytell about himself and build a community for other people who appreciate his general brand of wackiness. Aside from *Breaking Bad* fans, the blog is going to spur anyone with a taste for psychedelic rainbows and flying Pop Tarts to turn to his friends and tell them to watch. Their interest in Scheer's work and their attraction to him as a personality will probably follow him long beyond the *Breaking Bad* finale.

With this Tumblr campaign, Scheer is on his way to joining a class of A-list performers like Betty White and Louis C.K., whose savvy use of pop culture and technology helped them build their star power and propel their careers to new popular heights.

SMIRNOFF: Doing It All Wrong

Oh my God, why did you even bother, Smirnoff? This post shows that the brand has no clue how the Tumblr platform works.

★ **Inane text:** You tell fans, "Need drink ideas? Check out @SmirnoffUS on Twitter." Why should they? What have you offered in this post that would make any liquor connoisseur believe that Smirnoff has anything interesting to say?

★ **No link:** If the goal were to encourage Tumblr fans to start following Smirnoff on Twitter, wouldn't it have made sense to add a link taking them there? Consumers have the attention spans of mosquitoes—you have to do as much of the work for them as you can.

★ **Boring photo:** It's bad to use a still photo on a platform where you have the option of posting exciting, attention-getting animated GIFs. But Smirnoff could have redeemed itself had its creative team at least done something artistic with the photo, like Absolut did back in the 1990s. What value could they possibly bring to their consumers with a stock photo of a Smirnoff bottle? Even simply making the bottle move from side to side would have been more interesting than this.

FRESH AIR: Knows Its Audience

POSTED ON 10 APRIL, 2013 | 188 NOTES | PERMALINK | Reblogged from nightowlauthor

I was sick and out of the office most of last week and so here on the Tumblr we didn't properly mark the passing of screenwriter and novelist **Ruth Prawer Jhabvala**. Jhabvala is perhaps most well-known for the screenplays she did for Ismail Merchant and James Ivory, including those for **Room With A View** and Howard's End. She won Academy Awards for Best Adapted Screenplay for both. Jhabvala was 85.

The New York Times:

> Over four decades, beginning in 1963, Mrs. Jhabvala made 22 films with Mr. Merchant and Mr. Ivory, all examining culture in one way or another, often one that has vanished. …
>
> The casts were top-shelf and largely British, laden with stars like Maggie Smith, Anthony Hopkins, Emma Thompson, Daniel Day-Lewis, Helena Bonham Carter and Vanessa Redgrave. For "Mr. & Mrs. Bridge" (1990), based on the Evan S. Connell novels, Paul Newman and Joanne Woodward, spouses in real life, were recruited for the title roles.
>
> But Mrs. Jhabvala's writing was essential. She contributed sophisticated dialogue and a sharp eye for the nuances of class and ethnicity, Stephen Holden wrote in The New York Times.

Above, perhaps one of the greatest movie kisses all time from Room with A View and starring **Julian Sands** and **Helena Bonham Carter**.

RUTH PRAWER JHABVALA RIP OVERSIGHTS

For a staid media company, NPR has shown surprising and admirable savvy as it has successfully rebranded itself from a radio broadcaster to a disseminator of information and entertainment across all digital platforms. Its arts and culture talk show *Fresh Air* shows a similarly astute sensibility with this example of perfect Tumblr micro-content:

★ **Native art:** The only drawback of animated GIFs is that they don't translate well to the book page, so only by going directly to *Fresh Air*'s Tumblr blog will you be able to experience the full effect of the repeated loop of this scene from the Merchant-Ivory film *A Room with a View*, in which George, played by Julian Sands, passionately kisses Lucy, played by Helena Bonham Carter, looking sweet and innocent in her pre–Bellatrix Lestrange days. But it's totally worth going to the site to see *Fresh Air*'s perfect execution in their blog post commemorating the passing of screenplay writer Ruth Prawer Jhabvala.

★ **On-brand text:** Normally, this much text on Tumblr would be a turnoff, but this content was created for the NPR audience, and the NPR audience is made up of avid readers. It would have been out of character not to explain why the blog had previously ignored Jhabvala's death. In addition, the text is so personal and so *Fresh Air* you really get a sense of the human beings behind the blog.

ANGRY BIRDS: Playing Toward Emotional Investment

The Tumblr blog where this art appears was nominated for a 2012 Webby award. Rovio, the parent company that created a cultural touchstone with the video game Angry Birds, created the art. Angry Birds then combined it with another cultural touchstone, Star Wars, to create the mega-successful Angry Birds Star Wars. There are many reasons why the site is so popular, but there is one detail that warrants special attention, because it shows that the company really gets Tumblr:

★ **They invited the community in.** You'd expect the quality of any art put out by Rovio to be of the highest caliber. But if you look at the banner on the left side of the image, you'll see that Rovio didn't create this art at all. A fan made it. And Rovio has taken pains to make sure that everyone knows it. That's an incredibly smart move on the company's part. The Tumblr family is a deeply committed community, and Rovio wisely realized that if it invited followers to participate in the blog, and not just follow it, they would transfer much of their demo's emotional investment to the blog. It's an excellent way to build community and drive brand awareness.

LATE NIGHT WITH JIMMY FALLON:
Fanning the Flames of Greatness

Jimmy Fallon's Tumblelog, packed with reblogged material from fans creating animated GIFs with clips from his show, provides an excellent example of how to storytell on Tumblr. Post after post, we're entertained by the lunatic facial expressions and funny lines of his guests and fellow comedians like Amy Poehler and Retta Sirleaf. With this particular piece of microcontent, Fallon uses two animated GIFs as a gateway drug to the hard stuff, piquing our attention so that we're compelled to click on the link that takes us to YouTube, where we can watch the interview with Adam Scott in its entirety. This content succeeds on every level:

- **Makes use of content originally posted by a fan?** Check.
- **Acknowledges said fan so that other Tumblr users can find her?** Check.
- **Animated GIF?** Check.
- **Word-of-mouth-worthy?** Any Tumblr follower who just turned forty or knows someone who's about to could share this content, and gauging by the more than two thousand notes it earned, measuring how many times this GIF was liked or reblogged, it looks like they did.

AMAZON MP3: Throwing a Straight-Up Ask

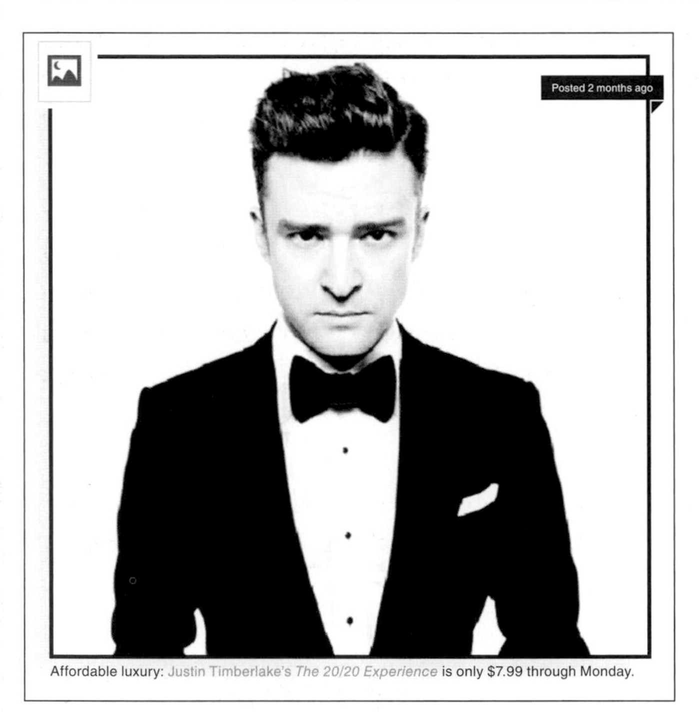

Affordable luxury: Justin Timberlake's *The 20/20 Experience* is only $7.99 through Monday.

I love this right hook mostly for the mere fact that it exists. It may bear the Amazon name, but the Amazon MP3 store does not have the brand awareness of its juggernaut parent company, placing it much closer in rank to an ordinary retail store. I get a lot of questions about how retailers should play on social, and this is a great example.

It's interesting to note how much black-and-white images overindex on Tumblr. Obviously the Amazon MP3 store is working with promotional materials for Justin Timberlake's album, so maybe this choice was just lucky. Regardless, the team knew well enough to take advantage of a striking, dramatic image.

★ **The copy is crisp and on voice for the audience and for the album:** Just two words—"affordable luxury"—make us feel like we're getting a premium product for a bargain. The link takes us straight to the product and store—no hunting around necessary. And finally, right there in the copy, the price—$7.99 through Monday. Nothing coy, nothing bashful. This is not a half-pregnant ask.

★ **This micro-content encapsulates the message of this entire book:** If you jab properly beforehand—through bringing your customers value in the form of a chuckle, or infotainment, or breaking news—you can say "Buy now!" and "Buy this!" without sounding like a carnival barker. Strong jabs buy you permission to throw unabashed right hooks.

WWF: Undermining Its Own Great Resources

There is a very small part of me that takes pleasure in critiquing this piece by the World Wildlife Fund. It's a little retribution for all the pain I suffered after the WWF forced the World Wrestling Federation, which also went by WWF, to change its name to World Wrestling Entertainment.

World Wildlife Fund has some gorgeous photographs on its blog. This picture of a man with a small child in his lap is one of them. Unfortunately, WWF has done nothing to make it memorable. There is nothing boring or dry about the issues championed by the WWF, and yet their Tumblr blog is about as inspiring as an empty sandbox. There is no story to grab our attention, no reason why we should stop to find out who is depicted in the photo, and no clear call to action.

- ★ **Dry, boring text:** "Just uploaded one new photo on Flickr." And??? Then, when we click to Flickr to see the photo, we're confronted with such dull copy it feels like it was cut and pasted from a database. There is no storytelling going on here.
- ★ **Weak call to action:** It's not until we click on the link to WWF's Flickr account that we learn that this is a picture of a community leader in Borneo and his five-year-old son. It informs us that this community is engaged in something called the Kutai Barat project, which "helps communities along the River Mahakam secure land tenure rights and livelihood skills." Then, the only additional link takes you back to the WWF home page, not to a page dedicated to the Kutai Barat project.

Just uploaded 1 new photo(s) on Flickr. http://bit.ly/PEoiEU

WWF has access to all the resources it needs to tell some of the most compelling stories on Tumblr, but here it missed the mark, and badly.

DENNY'S: Showing Some Delicious Moves

This is just one example of the tremendous work Denny's is doing on Tumblr.

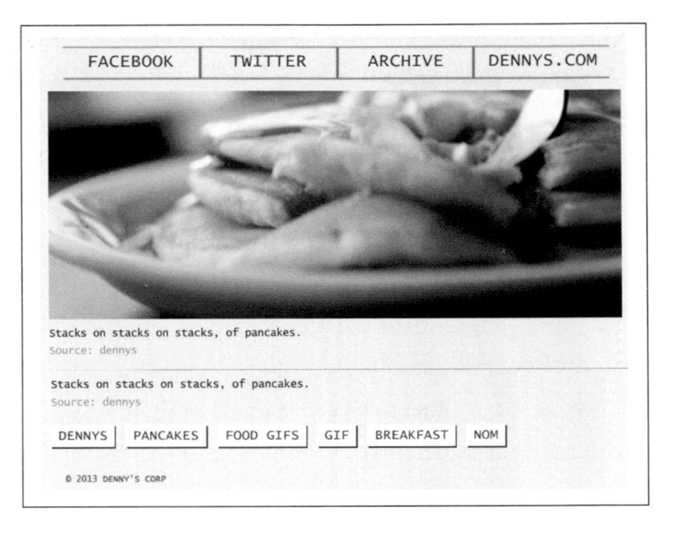

★ **Great GIF:** They are pros with animated GIFs. In this post, a fork repeatedly scoops up a soft heap of steaming hot pancake dripping with syrup.

★ **Great link:** Above, if you can tear your eyes off the GIF, you'll see four huge links to the company's Twitter feed, Facebook page, Tumblr archives, and corporate website. You can't miss them.

★ **Great text:** The text plays off the popular YC song "Racks," demonstrating that this brand, traditionally popular with families and retirees, knows how to talk to Millennials, too. So much so, in fact, that a blogger who goes by the name Synecdoche, a New York–based writer with a large Tumblr following, felt compelled to reblog this post to all of her followers. For a corporate brand to get praise from an anticorporate personality like her is like getting welcomed to the in crowd. It's the kind of word of mouth that has major impact on your business, the kind that can drive a car full of hungry rap-loving Tumblr users to pull into a Denny's parking lot.

TARGET: Hitting a Bull's-Eye

To see a pitch-perfect example of native storytelling and a strong right hook, take a look at this page on Target's aptly named Tumblr blog, On the Dot. It shows a dress. Specifically, a skater dress with a keyhole back. And in 3.7 seconds a flashing animated GIF allows us to see every version of it—black with studded collar detail, black and white stripes, bright floral, turquoise with white polka dots—while at the same time demonstrating the swishiness of the skirt.

★ **The piece has a clean look:** The animated GIF of the dress stands out against lots of white space and a bare minimum of elegant black script.

★ **Direct call to action:** Immediately below the GIF, three links (the polka dot dress is in-store only) let you pick the dress you want and take you straight to the Target website so you can buy it. The tags are perfect, too.

Somebody at Target knows exactly what he or she is doing.

GQ: Showing Mad Smart Tumblr Skills

To celebrate *Mad Men*'s sixth-season premiere, *GQ* announced "Happy Mad Men Day!" on Tumblr, accompanying the post with a photograph of many of the show's characters indulging in yet another cocktail hour. Here's why it scored more than two thousand notes:

- ★ **They paid attention to pop culture:** Millions of people were waiting with bated breath for the return of their favorite midcentury ad execs. *GQ* was smart to capitalize on their enthusiasm for the show.
- ★ **Smart links:** Not only is there a link below the photo, but the photo itself links to a meaty article, "The *GQ* Guide to Mad Men," which *GQ* published a year earlier on the eve of the show's fifth season, serving to remind followers where they can go to get more in-depth *Mad Men* coverage.
- ★ **Appropriate tagging:** Tagging is an extremely important part of Tumblr culture, and here *GQ* made smart use of it, including keywords like "Television," "John Slattery," "Jon Hamm," "Don Draper," and "Mad Men."

Questions to Ask About Your Tumblr Content

Did I customize my theme in a way that properly reflects my brand?

Did I make a cool animated GIF?

Did I make a cool animated GIF?

Did I make a cool animated GIF?

ROUND 8: OPPORTUNITIES in EMERGING NETWORKS

Every year, the world becomes a little smaller, a little more social, a little more connected. Creating content that allows us to share our experiences, thoughts, and ideas in real time is becoming an intrinsic part of life in the twenty-first century (in fact, it's getting to the point that we're making a statement when we *don't* share or choose not to connect). That's why it's smart to consider the jab-and-right-hook potential of platforms that aren't particularly social. It's just a matter of time before users adapt them, or demand that the developers adapt them, to provide the social layer people increasingly expect and crave. Whatever isn't a social experience now soon will be.

LinkedIn

- Launched: May 2003
- 200 million members
- Every second, two new members join.
- More than 2.8 million companies have a LinkedIn Company Page.
- Executives from all 2012 Fortune 500 companies are members.
- Students and recent college graduates are the site's fastest-growing demographic.

I predict that in the next twenty-four months, checking in to LinkedIn will be like checking in to Facebook—a regular part of our daily existence. Each social network will serve a distinct and vital purpose in our lives, like rooms in a virtual Downton Abbey. Facebook will be our dining room, where we entertain and get to know one another; LinkedIn will be our library, where we get deals done.

LinkedIn is already working hard to encourage more content creation and evolve from mere networking tool to professional hub. Users can now share articles, reviews, and examples of their work with their connections. The site has also launched the LinkedIn Influencers, where leaders contribute articles on their topic of expertise. All of this mirrors much of what can already be done on Facebook, and LinkedIn still has a long way to go to match the juggernaut's scale. But in one regard LinkedIn does have an advantage. As an exclusively business-oriented site, it provides a natural platform for B2B marketers, who have so far not seen much reason to bother with Facebook. If you're an office supply company, or a lawyer, LinkedIn can be an interesting place to tell your story, especially now while there is little to distract your fans. It is fertile jabbing ground for all business and brands, though, B2B or not. For more incentive, just imagine the spending power of the LinkedIn audience. LinkedIn's relevance hasn't reached a level where you need to post content at the same pace as you might on other social networking sites, but it would be wise to keep yourself in play here.

LinkedIn will be where you have the most freedom to indulge in long copy. Consider what people are looking for when coming to the site. They're hungry for information, they're looking for a job, they need to find an in or an edge, they want to meet professionally like-minded people. There have to be creative, smart ways you can make yourself indispensable to someone in that frame of mind. You can afford to be less flashy, maybe a little more serious and thoughtful. Or maybe not. Maybe you steer clear of slang and *OMG* and *LOL*, but you still add a welcome breath of levity to a serious environment. The key to giving your brand momentum on LinkedIn will be to offer native content that's completely different—and that has completely different value—from what you offer fans on other social networks.

Google+

- Launched: June 2011
- 500 million users

The future of Google+ as a viable marketing platform is one big question mark. Right now, Google+ is where Twitter was in 2006 or 2007. It has a big selling point: its implications for a Web property's SEO. Google gives preference to its own products, so having a Google+ account influences your search rankings. Still, so far it's really only the early tech influencers who are there, just like they were on Twitter in the early days. The platform hasn't caught on as quickly as Twitter, however, because there are more alternatives now than there were when Twitter came on the scene. Most of the world just isn't that interested in Google+ as a stand-alone product, because it offers little that people can't already get through Facebook.

The numbers say differently. Google+ points to its 500 million users as proof that it is steadily gaining a fan base. But the numbers are as overinflated as the lips of a Beverly Hills housewife, because Google requires consumers to sign up for a Google+ account if they want to register for any other of their product accounts, like YouTube. Look closely and you'll see that a large percentage of all those Google+ accounts are dormant. It is completely reliant on the scale and power of Google's other products.

But if Google Glass takes off in the way I think it will over the next half decade, Google+ will have a shot at rivaling Facebook for consumers' hearts. Why? Facebook and all the other social media platforms are scrambling to adapt for mobile. But Google Glass could conceivably replace mobile devices. It's going to allow users to record everything they see, and stream it live. It will be capable of putting a map directly in your line of vision, show you Google results on command, and it will be entirely voice-activated and hands-free. With that kind of technology, who's going to need a cell phone?

Now, there are two ways this could

go. Facebook will want to develop an app to let its users see what their friends are streaming on Glass, and Glass would understandably want to take advantage of Facebook's scale to build its user base. However, Google could also decide to make the product a closed network, requiring anyone who wants to see content on the glasses to log in through a Google+ account. If the glasses capture the public's imagination, and the only way to use it is through Google+, they're going to start spending a lot more time in those now-dormant accounts. As Google continues to natively integrate Google+ into all the other Google services and devices that people already love—Search, Gmail, YouTube, and Android devices—it'll be a knockout win for Google. And since the platform is so similar to Facebook, it won't require marketers to wrestle with reinventing their content strategy.

Vine

- Launched: January 2013
- As of June 2013, Vine gained 13 million users.
- In the week following its launch, almost half the videos posted on Twitter originated from Vine.
- Five Vine videos are shared every six seconds on Twitter.

Restrictions are a powerful thing. Although we often chafe at the limitations imposed by our marketing platforms, those limitations often bring out our storytelling creativity. That's why we should all be paying close attention to Vine, the six-second looping-video platform Twitter recently bought and launched to a whole lot of hoopla. By the time this book comes out, we're going to see how its restrictions have inspired some incredibly powerful storytelling. Currently, a lot of potential viewers pass over the opportunity to watch videos because they can't be sure if they're about to get sucked in for ten seconds or ten minutes, and that's not counting the preroll. The promise of a six-second limit is going to encourage a lot of people to watch Vine videos, representing a great opportunity for the marketer with the right skill set.

Truth be told, I'm infatuated with Vine. I think that six-second promise is going to turn it into one of the major platforms in the marketplace. It's the perfect product for our world—offering enough variety to satisfy the cravings of consumers constantly looking for their next dopamine hit, short enough for those time-pressed consumers to come back for more over and over again. One father I know told me that Vine was causing problems for his fifteen-year-old daughter, because she was staying up until three in the morning watching Vine videos. When asked why, she said it wasn't intentional. She'd decide to shut down, but then she'd spot a new video, and think, "Well, just one more—it's only six seconds."

Brands and businesses need to make it a priority to figure out Vine. Much like Instagram and Facebook before it, it is skewing young in its early months, appealing to eight- to twenty-one-year-olds. In twenty-four to thirty-six months, however, that demo will increase substantially, and businesses will need to be there. This platform could do to YouTube what Twitter did to Facebook. There will always be longer-form stories that will be better suited for YouTube, but Vine will become the video-watching platform of choice, especially because of its integration with Twitter. For further incentive, consider this: as of March 2013, consumers share branded Vine videos four times as often as branded Internet videos.

My only regret is that the platform just isn't mature enough for me to be able to tell you more about how best to use it. The best I can do is urge you to pay attention to how you edit your videos. A lot of people make the mistake of shooting an image for six straight seconds. That's boring. Just as edits and cuts are what build rhythm and suspense into a full-length movie, edits and cuts are critical to storytelling on Vine. There will probably be one or two major changes to the platform in the near future, but I, along with the rest of you, will be working my ass off to figure out how to best use this amazing tool as it evolves. I'm currently trying to create a new agency to represent the best Viners in the world. Check back with me to see if I pulled it off by the time this book goes into print.

Snapchat

- Launched: September 2011
- 60 million "snaps" are sent per day as of February 2013.
- My Snapchat name is GaryVayner.

Launched in 2011, Snapchat, the service that allows users to send photos and videos that self-destruct in a matter of seconds, was immediately labeled a sexting platform. Many people would be surprised to discover that it's actually used much more for circulating visual gags and jokes than dirty pictures. Snapchat was made for a world that can't stand one minute of boredom, and that is fast becoming addicted to putting out content. I share, therefore I am. Whereas before, the Internet operated under the 90-9-1 rule—the principle that generally, 90 percent of Internet users consume content, 9 percent edit it, and merely 1 percent create it—apps like Snapchat are going to shift those ratios to reflect something more along the lines of a 75-20-5 rule. Snapchat is not for profound content, nor to produce anything that's going to be treasured for eternity, or even analyzed as a case study one day. It's where people will go for a quick laugh before moving on. Imagine the power of a brand or business that can jab well enough to become the source of choice for those little moments that get us through our day. It's also a place where your content is likely to get more focused attention than on any other platform, because knowing that your content will disappear in seconds is incentive enough for consumers to make sure they don't miss it.

As usual, this new platform has been derided for its low value. It's not useful. No one will use it for anything important. It has no value. We've heard it all before. The debate swirling around the actual value of Snapchat is the same debate that was swirling around Facebook and Twitter not so long ago. Yet someone is clearly finding value in the platform when more than 60 million images are being sent around on a daily basis. And that value is only going to grow as the platform matures.

For now these platforms offer limited opportunities for right hooks. But for now isn't forever. Someone is going to figure out how to do it. It could be me. It might be someone else. Why not make it you?

ROUND 9:
EFFORT

Content is king, context is God, and then there's effort. Together, they are the holy trinity for winning on Facebook, Twitter, and any other platform, and even for winning in any business. Without effort—intense, consistent, committed, 24-7 effort—the best social media micro-content placed within the most appropriate context will go down as gracelessly as James "Buster" Douglas when he crashed to the mat at the end of his November 1990 fight with Evander "the Real Deal" Holyfield.

It's a sad story, though it should have been the next *Rocky*. At the time of the fight, Douglas had been enjoying fame as the world heavyweight champion after unexpectedly trouncing the then-undefeated heavyweight champion, "Iron" Mike Tyson, nine months earlier—an upset that put my fifteen-year-old self in such a state of shock, I hid in my bed and missed a day of school. Dead serious.

No one had expected Douglas to win that earlier fight. Tyson was the best boxer in the world; some thought he was the best boxer ever. This was the tenth time he was defending his title. Douglas had proved to be an unreliable fighter at best, and often carried more weight than he should. The odds were so high in Tyson's favor that only one casino would even take bets on the fight. Most people were watching just to see how fast Tyson could knock Douglas out.

But Douglas had done something no one expected of him—he trained like a man possessed. He was partly motivated by the unexpected death of his mother: "I knew that, somewhere, she was saying, 'That's my boy. He's gonna do it.' If I didn't do my best, if I didn't do what I was capable of, I thought my mama's ride to heaven would be a little harder. I didn't want that." But he had also met Mike Tyson in person and walked away unimpressed. Tyson couldn't be the undefeatable monster everyone said he was, and Douglas was going to prove it. By the time he stepped into the ring with Tyson, he had more than doubled his bench-press capability from 180 to 400 pounds, lost more than thirty pounds of weight, and watched countless tapes of Tyson fights. He studied Iron Mike's techniques, identified his flaws, and with the help of his managers and trainers, put together a strategy to take him down.

The effort paid off. Despite having been laid up with the flu just twenty-four hours earlier, Douglas pummeled Tyson with a series of strong, confident jabs, until at one point Tyson, his eye almost completely swollen shut, was literally using the ropes to keep himself standing upright. Douglas handed Tyson the first defeat of his career.

Effort is the great equalizer. It doesn't matter if your competitor is three times bigger than you and built like a Mack truck, or if it has a marketing budget that matches the GDP of a medium-size country, or if it has a staff of hundreds and you are alone in your broom closet with two laptops, an iPad, and a cell phone. What matters is the effort you put into your work. And never has effort counted more than it does today. Social media gave access to the market and even an edge over corporate behemoths to creative, determined, nimble upstarts. But now that big business has finally started investing in social media platforms like Facebook, albeit hesitantly, entrepreneurs no longer have as strong an advantage as they once did. One or two people just can't be in as many places at once building community as a staff of twenty. What they can still do, however, is win through effort. Budgets should have no effect on the amount of effort, heart, and sincerity that can go into your conversations with your customers. You can't be everywhere at once, but when the quality of your communication and community-building efforts is better than anyone else's, it doesn't really matter.

If you throw a great jab or right hook on Facebook, people will start to comment. Marketers that creatively *and* sincerely engage in as many of those resulting conversations as possible will be able to scale their relationships higher than their opponent. You should make sure to tag the person you want to talk to, to guarantee they see you have replied, and to bring them back to your page to continue the dialogue. Maybe you see that some people are unclear about the hours of your Black Friday sale, or they're not sure that the sale is happening in all your store locations. By going back and clearing up the confusion, you're amplifying your right hook and solidifying your relationship with your customers. And while you're there, be charming. Be funny. Show you care. People love to be entertained and informed, but they'll take that from anyone. The real connection, and the loyalty, happens when they believe that you care about them both as a customer and as an individual. People are usually astonished when a brand puts in extra effort to make them happy. That's how rarely it happens, and that's where you, whether you're an entrepreneur or a big business, can separate yourself from the rest of the pack. Bigger businesses will be able to jump in on more conversations than others, but volume alone won't raise a brand's engagement levels—the quality of the conversation will.

The thing to remember, however, is that you're fighting a never-ending boxing match. It's true that brands that throw out consistently skillful storytelling jabs and right hooks can eventually build up so much brand equity that they don't need to engage with quite the same frenzy as a newcomer, or a brand working to repair its reputation, but it's all relative. Twenty percent less of massive amounts of engagements is usually still more than most marketers' average engagement levels. But you cannot get lazy and rest on your laurels. You've got to keep putting in the effort, or you'll get knocked out in ten minutes. In Buster Douglas's case, it took seven minutes and forty-five seconds, to be exact.

Douglas's story, which started out as an underdog triumph, took a disappointing turn about nine months after his historic win over Mike Tyson. When he left the ring in February, he was in the best shape of his life and the new heavyweight champion of the world. He spent the next months doing the media circuit, appearing on the David Letterman show, posing for the cover of *Sports Illustrated*, signing autographs, and enjoying his fame. At the same time, however, he was still grieving for his mother. He also admitted to suffering from stress and depression due to a dispute with Don King, the boxing promoter with the electric shock of hair, who did his best to have the results of the Douglas-Tyson fight overturned. What he did not do was return to the training ring with the same intensity that he had when preparing for the Tyson fight. By the time he weighed in for his fight with Evander Holyfield, he looked like he'd eaten every cheeseburger in the world.

When Douglas and Holyfield faced off in the ring on November 9, 1990, they didn't seem overwhelmingly mismatched. Even the announcer commented, maybe with a little surprise, that there wasn't much difference in size between the two. What he didn't comment on, yet what was obvious the second each fighter took off his robe, was the difference in their shape. Holyfield's trapezii were so built up, his head appeared to be sitting on a perfectly sharp, muscular triangle. His massive shoulders and chest seemed cut from granite, beautifully defined as a statue's. As Douglas strutted into his corner, however, the tire around his waist jiggled slightly above his shiny white shorts; as he bobbed on his toes, his pecs shimmied, and his breasts drooped like soft little sponges. As the fight began, it was like watching a bull face off against a bulldog. Douglas was knocked out in the fourth round.

Effort. It matters more than most people want to admit.

ROUND 10: ALL COMPANIES ARE MEDIA COMPANIES

I've just spent nine chapters emphasizing that the key to social marketing is micro-content. In fact, the shorter your content and storytelling, the better. But as I look to the future, I see a yin to the micro-content yang. After all, long-form content isn't dead. It still lives on in the form of YouTube videos, magazine articles, TV shows, movies, and books, for instance, where it continues to find a sizable audience. But as brands continue to push the traditional boundaries via which they used to disseminate their content, and as companies recognize that they less and less often have to rent their media, but can own it and remarket it whenever they want, they're going to start wondering why they have to deal with separate media companies at all? Why couldn't they simply become their own media company? It's not a crazy idea. There is no logical reason to think that a tire company should be a food critic, but a hundred years ago, Michelin tires started reviewing rural restaurants to encourage people living in the cities to

drive farther and wear their tires out more quickly. Guinness created the *Guinness Book of World Records* to reinforce its brand and give people something to talk about in the pubs. Similarly, I predict that one day a brand like Nike could put out its own sports programming and compete successfully against ESPN, or Amtrak could launch a publication that could stand up to *Travel + Leisure*. The start-up costs would be extremely low for a luxury brand like Burberry to publish an alternative to the *Robb Report*, or for Williams-Sonoma to publish its own version of *Eater* or *Thrillist*. So long as brands remain transparent, so that their consumers aren't duped into thinking these sites and publications are strictly objective content providers, this could be a fruitful way to expand their brand and their content reach. In a way, it would be no different from what I was doing through Wine Library TV. Everyone knew that I sold wine, but they trusted my product reviews because I made a huge effort to be honest, fair, and authentic. Any other brand could do the same for the product or service they sell.

Some people will be skeptical. That's to be expected, especially in the older set. But the young, the under-thirty demo that puts a ton of stock in its bullshit detector? They know this is the future, and it doesn't scare them. They have come of age in the era of transparency, and they know they have no choice but to treat their consumers with honesty and respect. No consumer will put up with anything less.

In the marketing world, there will soon be no more separation between church and state. It's going to be exciting to witness the innovation that comes about as brands become major players in the media world.

ROUND 11: CONCLUSION

It takes a ton of effort to figure out how to use any social media platform to its full potential, and today we've got seven major ones to contend with. I hoped that if I wrote a short, useful book—one as visually enticing as a Tumblr or Pinterest post—that broke today's most popular and exciting platforms down to their essential building blocks of text, image, tone, and link capabilities, it would make the exploding social media scene seem a little less daunting to any marketer or business owner trying to keep up with it. I promise, the investment you make in familiarizing yourself with the ins and outs of these platforms will pay off, now and in the future. The rate at which they change is volatile, but the truth is that most companies and consumers are slower to adapt than they should be. This fact works in your favor. It means that you will have a significant business advantage if you choose to be part of the fraction of marketers who take the time to fully excavate these platforms' secrets. And it will be just a fraction. It always is. A member of the Google Analytics team recently informed me that almost no one uses the tracking system properly. Google Analytics has been around for eight years already, more than long enough for marketing departments to know it inside and out. But people's perception is that it's overwhelmingly complicated and vast, so even the best e-commerce companies have not put in the time and effort they should have to figure out how to take advantage of all the available features. There are a few marketers out there who have, though, and the data they access helps them beat their competition every day. They understood that whatever time they invested in learning Google Analytics would not be much compared with the huge returns the knowledge they gained could deliver. Marketers who put in the effort to really understand the nuances and subtleties of the platforms explored in this book can and will dominate. Yes, it will be frustrating when Facebook once again makes changes to its algorithm and newsfeeds, and Twitter and Pinterest will probably make tweaks and redesign. But if you don't give in to the frustration, and do persist in staying alert and figuring out how to use these changes to your advantage, you'll instantly be leagues ahead of most of the marketing pack. Others might huff and puff and eventually catch up, flattening your advantage, but you can make a lot of headway and be extremely effective during the two or three years that you're zooming ahead of the curve. Besides, if you've made staying ahead of the curve your standard procedure, what difference will it make if they catch up? To paraphrase Jay Z, you'll be on to the next one, probably to figuring out how to storytell on a Google Glass eye screen instead of a mobile phone. And about that time, I'll probably write a new book called *Four-Eyed Storytelling* or something like that.

In the meantime, when this book comes out I'll be storytelling all over the place, throwing jabs and right hooks at every opportunity. Maybe I'll put out a nine-second video on Facebook, then follow up with a short tweet accompanied by a link to Amazon. At the same time, you might find a picture of the jacket on Instagram, and then an animated GIF of the same picture doing wheelies on Tumblr. I'll have to figure it out. No matter what, though, I'll always be telling the same story—about social media, about business, and about how they really are now one and the same.

ROUND 12: KNOCKOUT

Just days before I was supposed to turn in the drop-dead, final version of the manuscript for this book to my editor, Instagram launched a fifteen-second video product that competes directly with Vine. I was in Cannes, and as soon as I could, I went back to my hotel room and spent four hours looking at every Instagram video I could find. And since then, my team at VaynerMedia and I, and all of the most progressive marketers in the world, have been scrambling to figure out the best way to storytell in fifteen seconds of video on a platform built for pictures. I can't think of a more fitting illustration of what kind of world we live in now.

Forget *Mad Men*, and fuck Don Draper. He lived in an easy world where nothing changed for thirty years, where you could spend your whole career working to figure out how the print and television markets worked. This world, the one you and I live in, evolves every second, every day. The skill sets it takes to be a successful entrepreneur, a successful marketer, or a relevant celebrity today is a different skill set than you needed ten years ago, even though that was the skill set that mattered for decades.

I have bad news: Marketing is hard, and it keeps getting harder. But there's no time to mourn the past or to feel sorry for ourselves, and there's no point in self-pity anyway. It is our job as modern-day storytellers to adjust to the realities of the marketplace, because it sure as hell isn't going to slow down for us.

Video for Instagram is just the most recent evolution. Soon Google Glass will launch and we'll have to figure out how to natively storytell on a screen hovering at the top of our customers' right or left eye. And as we go, we will have to continually reevaluate just how many times we should bring value through apps and videos and glasses before we can ask our consumers to do something for us. We have to remember to give, give, give before we ask. That will always be the real challenge. That, and moving fast enough to keep up.

The upside of moving quickly onto new platforms has been proved time and time again. The people and brands that over-index on Instagram and Pinterest are not necessarily the same ones who saw popularity on Facebook or Twitter—they just got to the new platforms first and figured them out sooner than anyone else. They're the ones that got out there and started testing, learning, and watching others. They went all in.

I hope you will, too. I hope you'll fight for your place in the social media ring with the same ferocity and conviction as Muhammad Ali and Joe Frazier during the Thrilla in Manila. If you don't know it, it's been described as one of the greatest boxing matches in history. Ali was officially declared the winner, but it's been said that both contenders fought so hard and so well that no one actually lost that day.

I like winning; I hope you do too!

NOTES

Round 1

1 there are nearly 325 million mobile subscriptions: Colin Knudson, "Smartphones, Tablets, and the Mobile Revolution," Mobile Marketer, January 29, 2013, http://www.mobilemarketer.com/cms/opinion/columns/14667.html.

2 almost half are networking on social media: "Americans Get Social on Their Phones," emarketer.com, August 8, 2012, http://www.emarketer.com/Article/Americans-Social-on-Their-Phones/1009247.

2 71 percent of people: Shayndi Raice, "Days of Wild User Growth Appear Over at Facebook," *Wall Street Journal*, June 11, 2012, http://online.wsj.com/article/SB10001424052702303296604577454970244896342.html.

2 almost half of all social network users: Andrew Eisner, "Is Social Media a New Addiction?" Retrevo Blog, March 15, 2010, http://www.retrevo.com/content/node/1324.

2 Boomers, who control 70 percent: Jack Loechner, "Booming Boomers," Media Post.com, August 21, 2012, http://www.mediapost.com/publications/article/181095/booming-boomers.html#axzz2XtXe5SNi.

2 Moms, the buyers and budget analysts: Melissa DeCesare, "Moms and Media 2012: The Connected Mom," Edison Research, May 8, 2012, http://www.edisonresearch.com/home/archives/2012/05/moms-and-media-2012-the-connected-mom.php.

4 It took thirty-eight years before 50 million people gained access to radios: United Nations Cyber Schoolbus, n.d., http://www.un.org/cyberschoolbus/briefing/technology/tech.pdf. Years it took the telephone to reach 50 million users: International Tele-

communication Union, "Challenges for the Network: Internet for Development," Executive Summary, October 1999, http://www.itu.int/itudoc/itu-d/indicato/59187.pdf. Years it took television to reach 50 million users: United Nations Cyber Schoolbus, n.d., http://www.un.org/cyberschoolbus/briefing/technology/tech.pdf. Years it took Internet to reach 50 million users: Ibid. Years it took Facebook to reach 50 million users: newsroom.facebook.com. Years it took Instagram to reach 50 million users: Chris Taylor, "Instagram Passes 50 Million Users, Adds 5 Million a Week," Mashable.com, April 30, 2012. http://mashable.com/2012/04/30/instagram-50-million-users.

6 you're investing only 1 percent of your ad budget: Kathryn Koegel, "Branding and Interactive Spending: Are We There Yet?," *Advertising Age*, October 29, 2012, http://adage.com/article/digital/branding-interactive-spending/238004/?utm_source=digital_email&utm_medium=newsletter&utm_campaign=adage.

Round 3

29 The platform was called: Sid Yadav, "Facebook, The Complete Biography," Mashable.com, August 25, 2006, http://mashable.com/2006/08/25/facebook-profile.

29 In a 2006 survey of: Mike Snider, "iPods Knock Over Beer Mugs," USAToday.com, June 7, 2006, http://usatoday30.usatoday.com/tech/news/2006-06-07-ipod-tops-beer_x.htm.

29 The "Like" button: Matt Lynley, "28 Crazy Facts You Didn't Know About Facebook," BusinessInsider.com, May 17, 2012, http://www.businessinsider.com/28-crazy-facts-you-didnt-know-about-facebook-2012-5?op=1.

29 Mark Zuckerberg initially rejected: Ibid.

29 There were more than a billion: Facebook Newsroom, Facebook.com, http://newsroom.fb.com/Key-Facts.

29 There were 680 million: Ibid.

29 One out of every five: Matt Tatham, "15 Stats About Facebook," Experian.com, May 16, 2012, http://www.experian.com/blogs/marketing-forward/2012/05/16/15-stats-about-facebook.

Round 4

83 As of December 2012: Tom Pick, "102 Compelling Social Media and Online Marketing Stats and Facts for 2012 (and 2013)," Business2Community, January 2, 2013, http://www.business2community.com/social-media/102-compelling-social-media-and-online-marketing-stats-and-facts-for-2012-and-2013-0367234.

83 The Twitter concept evolved: Eli Langer, "7 Things You Didn't Know About Twitter," BusinessInsider.com, March 17, 2013, http://www.businessinsider.com/7-things-you-didnt-know-about-twitter-2013-3.

83 The company's logo: Ibid.

83 JetBlue was one of the first: Andrew Moore, "A Conversation with Twitter Co-Founder Jack Dorsey," *Daily Anchor*, n.d., http://www.thedailyanchor.com/2009/02/12/a-conversation-with-twitter-co-founder-jack-dorsey.

83 Users post 750: Danny Brown, "52 Cool Facts and Stats About Social Media (2012 Edition)," Ragan's PR Daily, June 8, 2012, http://www.prdaily.com/Main/Articles/52_cool_facts_and_stats_about_social_media_2012_ed_11846.aspx#.

Round 5

117 48.7 million users: Craig Smith, "(June 2013) How Many People Use the Top Social Media, Apps, and Services?," Digital Marketing Ramblings, June 23, 2013, http://expandedramblings.com/index.php/resource-how-many-people-use-the-top-social-media.

117 From 2011 to 2012, Pinterest: Greg Finn, "Pinning the Competition: Pinterest's Four-Digit Growth is Tops in 2012," Marketing Land, December 4, 2012. http://marketingland.com/pinning-the-competition-pinterests-four-digit-growth-is-tops-of-2012-27769.

117 68%: "Pinning = Winning: The Infographic," Modea, February 25, 2012, http://www.modea.com/blog/pinterest-infographic.

117 The most repinned pin: Craig Smith, "Jabra Creates Contest to Find the Most Pinteresting Mom," Pinterest Insider, May 8, 2013, http://www.pinterestinsider.com.

117 the female demographic that outnumbers: Craig Kanalley, "Pinterest May Be Bigger Than You Think, Competing to Be the 2nd Most Popular Social Network,"

Huffington Post, February 15, 2013, http://www.huffingtonpost.com/craig-kanalley/pinterest-competing-twitter_b_2697791.html.

118 Pinterest was invented: Alyson Shontell, "Meet Ben Silberman, the Brilliant Young Co-Founder of Pinterest," Business Insider, May 13, 2012, http://www.businessinsider.com/pinterest–2012–3.

118 the approximately 48 million people: Sarah McBride, "Startup Pinterest Wins New Funding, $2.5 Billion Valuation," Reuters, February 20, 2013, http://www.reuters.com/article/2013/02/21/net-us-funding-pinterest-idUSBRE91K01R20130221.

118 That represents 16 percent: Maeve Duggan and Joanna Brenner, "The Demographics of Social Media Users, 2012," Pew Internet, February 14, 2013, http://pewinternet.org/Reports/2013/Social-media-users/The-State-of-Social-Media-Users/Overview.aspx.

118 Now that Pinterest has revised: Pinterest, http://about.pinterest.com/copyright.

119 A survey by Steelhouse: "Pinterest Users Nearly Twice as Likely to Purchase than Facebook Users, Steelhouse Survey Shows," Steelhouse press release, Steelhouse.com, May 30, 2012, http://www.steelhouse.com/press-center/pinterest-users-nearly-twice-as-likely-to-purchase-than-facebook-users-steelhouse-survey-shows.

119 Pinterest produces four times: "Advertising on Pinterest: A How-To Guide," Prestige Marketing, May 4, 2013, http://prestigemarketing.ca/blog/advertising-on-pinterest-a-how-to-guide-infographic.

119 Some small businesses: James Martin, "12 Things You Should Know About Pinterest," Life Reimagined for Work, January 23, 2013, http://workreimagined.aarp.org/2013/01/12-things-you-should-know-about-pinterest/#.UVH8nK3ERRM.email.

119 Between 2011 and 2012: Jeffrey Zwelling, "Pinterest Drives More Revenue per Click than Facebook," Venture Beat, April 9, 2012, http://venturebeat.com/2012/04/09/pinterest-drives-more-revenue-per-click-than-twitter-or-facebook.

120 doing so increases the number of likes: Mark Hayes, "How Pinterest Drives Ecommerce Sales," Shopify, May 2012, http://www.shopify.com/blog/6058268-how-pinterest-drives-ecommerce-sales#axzz2SEv3Ya59.

Round 6

135 As of December 2012: Greg Finn, "Pinning the Competition: Pinterest's Four-Digit Growth Is Tops of 2012," Marketing Land, December 12, 2012, http://marketingland.com/pinning-the-competition-pinterests-four-digit-growth-is-tops-of-2012-27769; http://www.theverge.com/2013/6/20/4448904/instagram-now-has-130-million-active-monthly-users.

135 40 million photos: Ibid.

135 It took Flickr two years: Mark Ashley-Wilson, "Some Fun Facts About Instagram #Infographic," Adverblog.com, August 18, 2011, http://www.adverblog.com/2011/08/18/some-fun-facts-about-instagram-infographic.

135 Instagram photos generate: Ibid.

135 Instagram started out: Kevin Systrom, "Instagram: What Is the Genesis of Instagram?," Quora.com, October 8, 2010, http://www.quora.com/Instagram/What-is-the-genesis-of-Instagram.

136 100 million monthly active users: Kevin Systrom, "Photoset," Instagram, February 2013, http://blog.instagram.com/post/44078783561/100-million.

136 With one new user: Katy Daniells, "Infographic: Instagram Statistics 2012," DigitalBuzz.com, May 13, 2012, http://www.digitalbuzzblog.com/infographic-instagram-stats.

Round 7

151 As of June 2013, 132 million: Tumblr Press Information, http://www.tumblr.com/press.

Hey, kudos on being an explorer and for finding this little Easter egg. Feel free to email me at gary@vaynermedia.com with any questions or thoughts about the book! Include the words *Easter egg* in your subject line so I know who you are!

151 60 million new posts: Tumblr Press Information, http://www.tumblr.com/press.

151 The Tumblr blog: David Karp, "Don't Laugh at Us," Tumblr.com, May 8, 2008, http://staff.tumblr.com/post/28221734/dont-laugh-at-us.

151 For every new feature: Liz Welch, "David Karp, the Nonconformist Who Built Tumblr," Inc.com, June 2011, http://www.inc.com/magazine/201106/the-way-i-work-david-karp-of-tumblr_pagen_2.html.

151 Ranks number one in average: Diana Cook, "Facebook's 900 million? But What About Engagement?" TheNextWeb.com, May 17, 2012, http://thenextweb.com/social media/2012/05/17/sure-Facebook-has-900-million-users-but-its-engagement-is -smoked-by-these-other-sites/?Fromat=all.

151 Bought for $1.1 billion: Chris Isidore, "Yahoo Buys Tumblr in 1.1 billion deal," CNN Money.com, May 20, 2013, http://money.cnn.com/2013/05/20/technology/yahoo -buys-tumblr/index.html.

152 He had tons of ideas: Tom Cheshire, "Tumbling on Success: How Tumblr's David Karp Built a £500 Million Empire," Wired.Co.UK, February 2, 2012, http://www .wired.co.uk/magazine/archive/2012/03/features/tumbling-on-success?page=all.

152 Tumblr's *obstsalat*: Ibid.

152 it debuted a newly streamlined dashboard: Sarah Perez, "With Today's Update, Tumblr Starts to Look More like a Fully Featured Twitter than Blogging Platform," TechCrunch.com, January 24, 2013, http://techcrunch.com/2013/01/24/with-todays -update-tumblr-starts-to-look-more-like-a-fully-featured-twitter-than-blogging -platform.

152 And in a *Forbes* interview: Jeff Bercovici, "Tumblr: David Karp's $800 Million Art Project," Forbes.com, January 2, 2013, http://www.forbes.com/sites/jeffbercovici /2013/01/02/tumblr-david-karps–800-million-art-project.

158 "If Leonardo da Vinci": Hugh Hart, "Animated GIFS Paint Breaking Bad Characters in Day Glo Pixels," Wired.com, April 12, 2012, http://www.wired.com/under wire/2012/04/breaking-bad-gifs.

161 going directly to *Fresh Air*'s Tumblr blog: Mel Kramer, *Fresh Air* on Tumblr, April 10 2013, http://nprfreshair.tumblr.com/post/47647361814/i-was-sick-and-out-of-the -office-most-of-last-week.

Round 8

172 200 million members: Jacco Valkenburg, "Everything You Want to Know About LinkedIn," Global Recruiting Roundtable, January 22, 2013, http://www.global recruitingroundtable.com/2013/01/22/linkedin-facts-figures–2013/?goback= .gde_52762_member_206908630#.UWcEfhnLInY.

172 Every second: Ibid.

172 More than 2.8 million: Ibid.

172 Executives from all 2012: LinkedIn Press Center, http://press.linkedin.com/About.

172 Students and recent college graduates: Montpellier PR, "25 Amazing LinkedIn Stats You Can't Miss," Montpellier Public Relations, January 17, 2013, http://montpellierpr.wordpress.com/2013/01/17/25-amazing-linkedin-stats-you-cant-miss.

174 As of June 2013, Vine gained: Jenna Wortham, "Vine, Twitter's New Video Tool, Hits 13 Million Users," *New York Times*, June 3, 2013, http://bits.blogs.nytimes.com/2013/06/03/vine-twitters-new-video-tool-hits-13-million-users.

174 In the week following its launch: "Jordan Crook, "One Week In, Vine Could Be Twice as Big as Socialcam," TechCrunch, January 31, 2013, http://techcrunch.com/2013/01/31/one-week-in-vine-could-be-twice-as-big-as-socialcam.

174 Five Vine videos are shared: Christopher Heine, "Twitter Vines Get Shared 4X More than Online Video: Researcher Says Nascent Tool Packs Branding Punch," *AdWeek*, May 9, 2013, http://www.adweek.com/news/technology/twitter-vines-get-shared–4x-more-online-video–149340.

175 consumers share branded Vine videos: Ibid.

176 60 million images: Jenna Wortham, "A Growing App Lets You See It, Then You Don't," *New York Times*, February 8, 2013, http://www.nytimes.com/2013/02/09/technology/snapchat-a-growing-app-lets-you-see-it-then-you-dont.html?_r=0.

Round 9

178 He was partly motivated: Robert Seltzer, "Fortitude Made Douglas a Big Hit, a Change of Heart Led to Triumph in Tokyo," *Philadelphia Inquirer*, February 15, 1990.

179 He also admitted to suffering: Author Unknown, "Douglas Weighs In at 246 vs. Holyfield," *Daily Record*, October 25, 1990, http://news.google.com/newspapers?nid=860&dat=19901025&id=4HhUAAAAIBAJ&sjid=e48DAAAAIBAJ&pg=6802,7359288.

ABOUT THE AUTHOR

Gary Vaynerchuk is first and foremost a storytelling entrepreneur. He is also a *New York Times* bestselling author, and his digital consulting agency, VaynerMedia, works with Fortune 500 companies to develop digital and social media strategies and content. *Businessweek* selected him as one of the top twenty people every entrepreneur should follow, and CNN voted him one of the top twenty-five tech investors on Twitter. He lives in New York City, where he avidly roots for the New York Jets.